Sissi Ram

Life in the Universe

Sissi Ram

Life in the Universe
Interstellar contact and diplomacy in space

Paperback

All rights reserved by publisher

Copyright © 2024

Imprint Page: 250

ISBN: 9789403788050

Dedicated to the pioneers who take up the challenge of exploring the unknown and building a bridge between our world and

"Life in the universe:
Interstellar contact and diplomacy in space"

Content

Foreword 9

Introduction 11

Significance of extraterrestrial contact 14
Historical perspectives 14
The challenges and opportunities 19
Target group of the guide 21

Foundations of Communication 25
Different Life Forms 25
Physiological Properties 29
Biological Properties 31
Psychological characteristics 34
Common communication foundations 39

Preparation for extraterrestrial contact 43
Possible forms of contact 43
Fears and prejudices 45
Mental and emotional preparation 48
Openness and respect 50

Communication strategies 54
Basic knowledge of diplomacy 54
Challenges 56
Goals and interests 57
Communication techniques and principles 61
Non-verbal communication, body language 64
Exchange options 70
Practical applications 73
Existing language for aliens 76

Challenges and solutions 78
Different scenarios 78
Effects of possible contact 79
Shaping relationships 81
Benefits of collaboration 82
Possible challenges and risks 84
Misunderstandings and conflicts 86
Building trust and establishing cooperation 89

Mental and emotional stability · · · · · · · · · · · · · · · · 93
Different techniques · 93
Insecurities and fears · 96
Empathy and understanding · 99
Insecurities and fears · 102
Empathy and understanding · · · · · · · · · · · · · · · · · · · 105

Look into the future · 109
Developments · 109
Significance for humanity · 114
Approaches and projects · 116

Framework conditions · 120
General legal aspects · 120
Rights and duties in the universe · · · · · · · · · · · · · · 122
Accountability at first contact · · · · · · · · · · · · · · · · · 123
Property rights · 125
Trade relations · 128
Living together in settlements · · · · · · · · · · · · · · · · · 130
Sovereignty · 132
Conflict resolutions · 133
Technology transfer · 135
Responsibility · 136
Ethics · 137
Spiritual and philosophical reflections · · · · · · · · · 139

Living with aliens · 142
Living together · 142
Intergalactic coexistence · 143
Space colonies · 145
Challenges and opportunities · · · · · · · · · · · · · · · · · 147
Diseases and viruses · 149

Extraterrestrial visits · 151
Speculation and discussions · · · · · · · · · · · · · · · · · · · 151

Eyewitnesses, UFO sightings · · · · · · · · · · · · · · · 154
The Project Blue Book (1952-1969) · · · · · · · · · · · · 154
The Majestic Twelve (formed 1947) · · · · · · · · · · · · 156
Who are the MJ-12? · 159
Kecksburg UFO incident (1965) · · · · · · · · · · · · · · · 164
Rudloe Manow (1974) · 166
Malmstrong nuclear missile crisis (1967) · · · · · · · 168
Roswell Incident (1947) · 171

Phoenix Lights (1997) 175
Vorfall in Rendlesham Forest (1980) 177
Incident of Ariel (1978) 180
Varginha incident (1996) 182
Travis Walton Incident (1975) 184
Betty and Barney Hill (1961) 188
Shag Harbor incident (Canada, 1967) 191
Cash-Landrum incident (USA, 1980) 194
Vostok Incident (Russia, 1982) 197
Río Cuarto (Argentina, 1965) 200
Hessdalen lights (1983) 203
AATIP 206

Your own journey with aliens 209
Exchange with like-minded people 209

Interesting facts 212
Five recognized UFO signs 212
UFO sightings and nuclear weapons 213
Paul Hellyer 215
Haim Eshed 216
David Marler 218
Lou Elizondo 220
Chris Mellon 222
John Callahan 224
Linda Moulton Howe 226
Bob Lazar 228
David Grusch US Secret Service employee 230
Senator Harry Reid 231
General thoughts 234

Epilogue 239

Sources 242

Disclaimer 250

Foreword

Welcome to our extraordinary guide. We invite you to join us in exploring a question that has always captured the human imagination: the possibility of the existence of extraterrestrial life forms and our potential interaction with them. Throughout human history, we have repeatedly looked to the heavens and wondered whether we are alone in the universe. There is growing evidence that we may not be alone, that there may be other intelligent beings in the vastness of the cosmos who, like us, seek contact and exchange.

This guide is our call for reflection and open discussion. It encourages us to think outside the box, to consider the possibility of the existence of extraterrestrial intelligences and to prepare for the potential challenges and opportunities that an encounter with extraterrestrials could bring. Here we will ask the question of how we might behave as individuals, as a society and as humanity as a whole if we find ourselves in the unexpected situation of having contact with extraterrestrial beings. We will look at different scenarios, examine alleged encounters and analyze the reactions and behavior of governments and institutions.

At a time when technology and science are advancing rapidly, it is crucial to consider the ethical, social and cultural implications of extraterrestrial encounters. This guide encourages you to prepare for possible scenarios and to work together to develop a constructive approach to enable harmonious coexistence with extraterrestrial civilizations. We hope that this book will arouse your interest, make you think and offer you suggestions on how we as a society could deal with a possible encounter with extraterrestrial life. Because only through open discussion and a willingness to consider new perspectives can we collectively prepare for a future that may extend beyond the boundaries of our planet.

We are delighted that you have embarked on this fascinating journey of exploration and imagination, that you have chosen "Life in the universe: Interstellar contact and diplomacy in space"

Sincerely
 Sissi Ram & Team

Introduction

Dear readers!

For centuries, mankind has been preoccupied with the question of extraterrestrial life. With our growing knowledge of the universe, it is becoming increasingly likely that intelligent life forms exist somewhere "out there".

But what should we do if one day we do come across extraterrestrials? How can we build a positive and productive relationship with them?

This book, entitled "Life in the universe: Interstellar contact and diplomacy in space", takes you on a fascinating journey as we explore various aspects of extraterrestrial contact. We explore the opportunities and challenges that an encounter with extraterrestrial life can bring. We cover topics such as preparation, different types of contact, possible diplomatic relations and the future of humanity in the context of extraterrestrial life.

This guide serves as a guide to prepare humanity for a possible encounter with extraterrestrial life and to minimize risks and challenges. We are increasingly confronted with news of unexplained phenomena and unknown flying objects. The opinion that extraterrestrials have been living among us in some form for a long time is also becoming increasingly common. Have we already thought about how we should behave in such a situation?

"Life in the universe: Interstellar contact and diplomacy in space" sheds light on legal, ethical, philosophical and spiritual issues that could arise from such an event from the perspective of an ordinary person. We may not present scientific results, as these are not accessible, but we offer you space to expand your own imagination and understanding. An encounter of the third kind could offer exciting and fascinating prospects, but we must not forget the challenges and risks. So why not try to prepare ourselves to achieve a high level of cooperation and understanding? This could ensure possible coexistence or cooperation with extraterrestrial species.

Such an encounter could provide humans with a unique opportunity to broaden their horizons and take knowledge and technology to a new level. It is our responsibility to ensure that we manage such an encounter peacefully and sustainably by respecting the rights and interests of all species involved.

With "Life in the universe: Interstellar contact and diplomacy in space" we would like to give you food for thought to reflect on these important issues and possibly prepare for an encounter with extraterrestrial life.

Please note that the guide serves as a guide and does not provide definitive answers to the questions of extraterrestrial contact. The future of extraterrestrial contact remains a fascinating and uncertain area of ongoing research and debate.

Thank you for your interest and good luck in your exploration of galactic communication!

Significance of extraterrestrial contact

Historical perspectives

Right at the beginning, we would like to address the question of why extraterrestrial contact is of great interest to humans? We have always been interested in the idea of extraterrestrial life and are fascinated by the idea that we are not alone in the universe. The reasons for this interest are manifold and range from the search for new knowledge to deeper existential questions.

Human nature is characterized by curiosity and the desire for discovery. The possibility of encountering extraterrestrial life is one of the most fascinating and profound discoveries we wish to explore. The idea that there may be intelligent life forms on other planets or in other galaxies awakens our spirit of discovery and drives us to seek new knowledge. Extraterrestrial contact may promise answers to fundamental questions about the universe and our own existence.

Where do we come from?
Are we alone in the universe?
Are there other intelligent civilizations?

The search for extraterrestrial life can help us to better understand our place in the universe and broaden our perspective on life and existence.

The idea of extraterrestrial life also has implications for our technological progress. If it is possible to make contact with an advanced extraterrestrial civilization, we could benefit from their knowledge and technologies. Extraterrestrial contact could help us to further develop our own technological capabilities and find new solutions to global challenges.

Contact with extraterrestrial life forms would have a profound impact on our world view. It would force us to rethink our ideas about the uniqueness of Earth and human life. We would have to remember that we are part of a much larger and more diverse universe in which there are countless possibilities for life and development. Extraterrestrial contact could also enable cultural exchange and cooperation between different intelligent civilizations. We would have the chance to learn from other cultures, gain new perspectives and work together on global challenges. Communication with extraterrestrials could lead to an enrichment of our own culture and society.

The idea of extraterrestrial life could also challenge religious and philosophical beliefs. Extraterrestrial contact could lead us to rethink our religious beliefs and seek new answers to theological questions. It could also trigger philosophical discussions on topics such as identity, morality and the purpose of life.

A visitor from space would potentially bring about profound societal changes. The discovery of an extraterrestrial civilization would expand our notions of nationhood, race and culture. It could lead to greater global cooperation to address common challenges and promote peaceful coexistence between different intelligent species. It would be important that we respect the habitats and rights of other species to enable harmonious coexistence. How can we ensure that communication takes place on an equal footing and that we learn from them without violating their rights? These questions require careful consideration and an ethical framework to ensure that extraterrestrial contact is positive and respectful for all involved.

Interaction with extraterrestrial life forms would also pose technological and scientific challenges. Communicating with a completely different intelligence may require the development of new communication methods and technologies. In addition, we would also have to consider the impact of extraterrestrial contact on our own technological and scientific advances. It could produce new insights and innovations in various fields such as astronomy, biology and physics.

It is also crucial that we consciously consider the potential impact of extraterrestrial contact and ensure that we deal with it responsibly should it occur. Exploring extraterrestrial life and preparing for contact requires extensive discussion and cooperation at a global level.

International cooperation would be of great importance in this context. Extraterrestrial contact would be a global issue requiring cooperation between all, or the majority of countries. It would also be important for governments, scientists and organizations to share resources and information to work together on exploration and communication with extraterrestrial life forms. International agreements and protocols could be developed to ensure the sharing of knowledge and the protection of all parties involved.

Furthermore, it would be crucial that we prepare for different scenarios when it comes to extraterrestrial contact. It is possible that we may encounter intelligent and friendly life forms, but there is also the possibility of encountering a completely unknown and potentially hostile species. We should prepare to develop protocols and strategies to deal with different situations to ensure we can respond appropriately.

Extraterrestrial contact could also have an impact on our natural environment. For example, if we receive information about other habitable planets, we could have the opportunity to expand our search for new habitats. This could lead us to become more concerned with the protection and preservation of our own environment to ensure a sustainable future.

Extraterrestrial contact would undoubtedly have a major impact on popular culture. Books, movies, games and other media would deal with the topic of extraterrestrial life and contact. It would spawn new stories, myths and legends and further fuel our imaginations. Popular culture could also serve as a means of making information and ideas about extraterrestrial contact accessible to a wide audience.

We can easily say that extraterrestrial contact is a topic that challenges and inspires us as a society. It opens up new perspectives, raises questions about our own existence and inspires deep reflection. By addressing these questions and preparing for possible scenarios, we can ensure that we can respond appropriately and benefit from extraterrestrial contact should it occur in the future.

Exploring extraterrestrial life and preparing for contact requires extensive discussion and cooperation on a global scale. It is vital that governments, scientists and society as a whole work closely together to understand the potential implications of extraterrestrial contact and jointly address ethical, technological and social challenges. Through international cooperation, resources and information can be shared to advance exploration and communication with extraterrestrial life forms. International agreements and protocols could be developed to promote the sharing of knowledge and ensure the protection of all parties involved.

In addition, we should prepare for various scenarios in order to respond appropriately to extraterrestrial contact. This includes developing new communication methods and technologies, as well as considering the impact on our environment and protecting our own natural resources. Extraterrestrial contact would undoubtedly also influence popular culture and inspire new creative works. Books, movies, games and other media have already and will continue to address the topic of extraterrestrial life and contact. This could help educate and raise public awareness.

The idea of visitors from outer space is a fascinating topic that makes us think about our existence, our place in the universe and the ethical and technological challenges that come with it. We should address these questions and begin preparing for possible extraterrestrial contact as a global community.

The challenges and opportunities

This chapter examines the challenges and opportunities that may be associated with extraterrestrial contact. Possible linguistic, cultural and technological barriers are highlighted and the impact these could have on communication is discussed. At the same time, the potential opportunities, such as the exchange of knowledge and perspectives, the promotion of intercultural understanding and possible joint projects are discussed. Extraterrestrial contact presents us with a number of challenges, but also holds fascinating opportunities. Linguistic challenges may arise, as it is unlikely that extraterrestrials speak or understand our human languages. Therefore, it requires intensive research and analysis to find a common basis for communication.

Another aspect that needs to be considered is cultural differences. Extraterrestrial life forms could have a completely different perception of the world and possess different value systems. This requires us to be prepared to question our own cultural perceptions and engage with new ways of thinking.

Technological barriers could also arise, as communication over long distances in space could require advanced technologies. The development and application of new communication technologies, such as the use of quantum entanglement or other theoretical concepts, may be required to enable effective communication. Contact from space also requires trust building and ethical awareness. It is important to adopt a respectful and responsible attitude to avoid possible negative effects on both sides. The establishment of ethical guidelines and the continuous reflection of our own intentions and actions are of central importance here.

Despite these challenges, extraterrestrial contact also offers immense opportunities and potential. The exchange of knowledge and information with an intelligent extraterrestrial civilization could greatly advance our own development. We could benefit from their technology, their science and their insights and find new solutions to global challenges. Contact with extraterrestrial life forms would help us to expand our consciousness and gain new perspectives. We could re-evaluate our ideas about life and our existence in the universe and expand our world view.

This increase in awareness could lead to a deeper understanding of the diversity and richness of life.

Through contact with different extraterrestrial civilizations, we could learn about different cultural perspectives and develop a deeper understanding of the diversity and richness of life. This could lead to greater tolerance, openness and cooperation between different cultures and societies. Extraterrestrial contact could also create the basis for joint projects and cooperation. By sharing resources, knowledge and technologies, we could work together on global challenges such as climate change, sustainable development or space exploration. This cooperation would have the potential to pave new paths of progress and peace.

However, always keep in mind that extraterrestrial contact is purely speculative so far and there is no scientific evidence for the existence of extraterrestrial life. Nevertheless, we should consider the potential challenges and opportunities of this contact. It opens up the possibility of broadening our perspectives, finding new ways of sharing knowledge and working together, and better understanding our place in the universe.

These ideas are part of a fascinating and speculative discourse that inspires our imagination and makes us think about our own existence in the universe.

Target group of the guide

In this guide, we want to ensure that all people, regardless of their scientific or technical background, have the opportunity to explore the topic of extraterrestrial contact. Our aim is to provide you with practical approaches and general knowledge that can help you in the event of a potential encounter with extraterrestrials.

Our guide does not require extensive scientific knowledge or technical expertise. We have made a conscious effort to make the content understandable and accessible so that any reader, regardless of their education or professional experience, can benefit from the information. Perhaps you are wondering why people with no scientific or technical knowledge should be interested in such a topic at all? The answer lies in our innate thirst for knowledge and our natural curiosity. Extraterrestrial contact is a fascinating subject that encourages us to use our imagination and think beyond our own limitations. It offers us the opportunity to look beyond the known boundaries of humanity as a whole and broaden our perspectives.

Our guide will help you develop basic communication strategies and strengthen your skills in dealing with potential extraterrestrial life forms. We will show you how to use nonverbal communication and body language to make yourself understood, and how to develop an open and respectful attitude towards other cultures and ways of thinking. We will also cover simple exchange possibilities and understanding of different thought patterns. We want to show you that it is not necessarily necessary to speak a common language. There are universal communication patterns and fundamentals that can help us to exchange with other intelligent life forms.

Please note that our guide cannot provide a comprehensive scientific overview of the topic of alien contact. Rather, it is a practical guide that gives you tools and strategies to prepare for a possible encounter. Our hope is that every reader, regardless of their background, will be inspired by this guide to begin their own journey of exploration and communication with extraterrestrials.

We are convinced that every human being has the ability to engage in this fascinating possibility to benefit from it. By addressing people without scientific or technical knowledge related to the exploration of the universe, we want to ensure that the topic of alien contact is accessible to everyone.

In our guide, we will try to introduce you to practical exercises and techniques that will help improve your communication skills. You will learn how to recognize signals and signs that may come from extraterrestrial life forms, and how to respond appropriately. Additionally, we will address the topic of intercultural communication, as we can assume that extraterrestrials could come from different cultures and ways of thinking. We will present you with strategies to avoid misunderstandings and promote harmonious and respectful communication.

Furthermore, we will address ethical aspects of alien contact. It is important to us that you maintain your values and principles and act respectfully in a possible encounter with extraterrestrials. We will provide you with guidelines and recommendations on how to make ethical decisions while keeping the well-being of all parties involved in mind. We want to emphasize that this guide does not claim to answer all questions about alien contact. The exploration and understanding of the universe is an ongoing process, and there are still many unresolved mysteries.

However, we hope that this guide will help you prepare for a potential encounter with extraterrestrials and improve your communication skills.

We want to encourage you to remain open and foster your own curiosity. Alien contact can be a unique and transformative experience. By developing our communication skills and being willing to engage with the unknown, we can not only advance our own development but also contribute to positive change on a global scale.

Foundations of Communication

Different Life Forms

To better understand the variants of communication with extraterrestrials, it is of great importance to comprehend the various known and potential life forms. Our current understanding of life is based on the assumptions we know from Earth. It is assumed that these conditions also apply to most terrestrial life forms. However, concepts of extremophilic life, which can exist under extreme conditions and may have different requirements, also exist.

In this guide, we delve deeply into the topic of communication with extraterrestrials. Naturally, the question arises as to what kind of life forms these could be. When considering Earth as a starting point, certain essential prerequisites are necessary for life to occur. For instance, water is a crucial component of life, as it is essential for biochemical reactions and the metabolism of vital organisms. Carbon, on the other hand, serves as the basic building block for living organisms, forming the backbone of their molecules and structures. Another essential component for life, both on Earth and potentially for extraterrestrial life forms, is oxygen. Oxygen plays a vital role in energy production through cellular respiration and is indispensable for many organisms. Organisms take in oxygen through respiration and release carbon dioxide in the process.

Furthermore, a protective atmosphere is of great significance as it shields against harmful radiation from space and creates a stable environment in which life can develop and thrive. Earth's atmosphere primarily consists of nitrogen, oxygen, argon, and other trace gases. Oxygen constitutes a significant portion of the atmosphere, enabling the survival of many living organisms. The presence of oxygen in the atmosphere plays a crucial role in the process of respiration, essential for many higher organisms. While plants produce oxygen through photosynthesis, animals take in oxygen to gain energy through cellular respiration.

Moreover, the atmosphere regulates temperatures on Earth by absorbing solar radiation and radiating heat back into space. This interplay of radiation and atmosphere creates climatic conditions crucial for the existence and diversity of life on our planet. In addition to the aforementioned main components, the atmosphere also contains hydrogen, helium, and small amounts of other gases.

The specific composition of the atmosphere acts as a protective shield against harmful influences from space, including intense solar wind and cosmic radiation. This protective envelope enables the Earth's surface to provide an environment in which complex and diverse life can develop and thrive.

When considering potential extraterrestrial life forms, it is crucial to examine whether and how their respective planets exhibit similar protective mechanisms against the extreme conditions of space.

These fundamental components-water, carbon, oxygen, and a protective atmosphere-form the basis for life on Earth and could also be relevant when considering extraterrestrial life. It is possible that extraterrestrial life forms have similar dependencies and requirements on their environment to exist. Therefore, it is important to consider these basic prerequisites when contemplating the possibility of communication with extraterrestrials.

However, it is also conceivable that there are other life forms that require different foundations for their existence. In our infinite universe, there exists an unimaginably large number of planets where life could potentially exist. But what exactly do we understand by "life", and what forms of it are possible? This chapter addresses precisely this question.

On Earth, we are accustomed to a limited number of living beings that we can perceive with our senses. However, there is also invisible life that only becomes visible under the microscope, such as bacteria, viruses, and fungi. These microscopic organisms represent a variety of life forms that are unknown to us. In addition to the familiar life forms, there is an incredible diversity of animal life. From tiny insects and spiders to majestic whales and elephants, animals exist in all possible forms and sizes. There are animals that live in water, those that fly in the air, and those adapted to life on land. The diversity of animal life demonstrates that there is a wide range of possibilities. Plants also play a vital role in life on Earth. Plants are able to derive energy from sunlight and water through photosynthesis. They form the basis of many ecosystems on Earth and provide food and habitat for numerous animals. The diversity of plants shows that there are various ways in which life on other worlds could appear.

Now, the question arises: What about extraterrestrial life forms? How could they look and function? There are numerous theories and speculations on this topic. Some scientists believe that there could be life on other planets that has evolved similarly to that on Earth. This view is based on the assumption that the basic building blocks of life, such as carbon and water, are also present elsewhere in the universe.

However, others think that there could be completely unimaginable forms of life that obtain energy and survive in other ways. The possibilities are limitless. Most known life forms on Earth are based on carbon, but it is also possible that other elements could serve as the basis for life. Life has been found in some extreme environments that is based on sulfur or other chemical elements. This demonstrates that there could be various chemical foundations on which life can build. Another possibility is that extraterrestrial life forms do not exist based on chemistry but could rely on fundamentally different physics and biology. Our conception of life is heavily influenced by conditions on Earth, but in the deep universe, entirely new forms of life could exist, functioning in ways entirely unexplored to us.

For example, some theories suggest that there could be organic life forms based on silicon instead of carbon. Silicon possesses similar chemical properties to carbon and could thus serve as the basis for alternative biochemical processes. Such silicon-based life forms could exist in extreme environments where carbon-based life could not survive.

Physiological Properties

The diversity of physiological properties of extraterrestrial life forms opens up a fascinating world of possibilities. When envisioning encountering such a species, questions arise about how they breathe, how their bodies are structured, how they feed, reproduce, and perceive their environment.

The respiratory system of extraterrestrial life forms could be entirely different from ours. While we inhale oxygen, these beings may require nitrogen or methane as vital gases, or they may have developed an entirely different method of oxygen uptake that is foreign to us.

The body structure could also vary remarkably. With a multitude of limbs, eyes, ears, and mouth openings, specifically adapted to the requirements of their environment and way of life, their sizes and shapes could reveal a new dimension of biological diversity. The digestive system could also function fundamentally differently, with food sources and energy acquisition being completely different, perhaps even without a conventional digestive system.

In reproduction, extraterrestrial life forms could exhibit a fascinating range of reproductive mechanisms, from spores to budding to division. Complex mating rituals and a variety of genders could characterize their reproduction, offering us a new perspective on the miracle of life. Additionally, their sensory organs could enable a completely different type of perception, such as infrared or ultrasound vision.

However, these examples only offer a small glimpse into the possible physiological properties of extraterrestrial life forms, which unfortunately remain unknown. It is of great importance to prepare for possible physiological differences, as our knowledge is currently based on speculative theories and fictional depictions. Exploration requires open and flexible thinking, a willingness to challenge existing ideas, and adaptation to the unknown.

We should be aware that the actual physiological properties of extraterrestrial life forms, if they exist, may be completely beyond our imagination. It is important to be prepared to encounter phenomena that challenge our current scientific models and explanations.

In this context, it is useful to further explore the limits of our own physiological experiences. By studying extremophiles on Earth and promoting interdisciplinary collaboration, we can better understand possible adaptations and survival strategies of extraterrestrial life forms. The physiological properties of extraterrestrial life forms could hold a world full of surprises and wonders, and it remains exciting to prepare for these possibilities, to maintain curiosity, and to open ourselves to the unknown. For perhaps the day will come when we actually encounter an extraterrestrial life form and can behold the physiological diversity of the universe with our own eyes.

Biological Properties

Let's assume we witness an extraordinary discovery – an encounter with extraterrestrial life. The excitement and curiosity would be unparalleled. As we embrace the unknown, it would be crucial to also consider the biological properties of these alien life forms.

The first fascinating aspect we might contemplate is metabolism. Extraterrestrial life forms could possess a completely different metabolism than the life we know on Earth. Perhaps they utilize an energy source that is unfamiliar to us, or they have an astonishingly efficient metabolism that allows them to thrive on minimal sustenance.

Another important aspect is reproduction. There are countless ways in which extraterrestrial life could multiply. Some might prefer asexual reproduction, in which they reproduce without a partner, while others may practice sexual reproduction, in which there is a union of sex cells. But beyond that, extraterrestrial life forms could also use forms of reproduction that are completely alien to us. Reproduction can not only be a biological aspect, but can also have psychological effects, such as mating behavior, mate choice, or the development of parental behavior. Thus, in the study of the reproduction of extraterrestrial life, biological and psychological aspects could overlap.

Another aspect that challenges our imagination is lifespan. The lifespan of extraterrestrial life forms could significantly differ from ours. Some may exist only for a short period, while others may endure for thousands or even millions of years. The concept of time and aging could unfold in an entirely new dimension. A remarkable characteristic of earthly life is its adaptability to various environments. Therefore, it is entirely possible that extraterrestrial life also possesses this ability and can thrive in extreme environments that would be inhospitable to terrestrial life. We may encounter life forms that thrive in extreme temperatures, under high pressure, or even in the vacuum of space.

The senses of extraterrestrial life forms could also hold a world full of surprises. They may possess senses that are entirely unknown to us or have developed a type of sensory perception to perceive their environment in ways we can hardly imagine.

We must acknowledge that our knowledge of the biological properties of extraterrestrial life forms is extremely limited. Our experiences are based solely on the life we know on Earth. Therefore, we can only speculate and hypothesize about what forms of life could exist in the universe. The true diversity and nature of extraterrestrial life, if it exists at all, ultimately remain hidden from us.

Despite these limitations, it is crucial that we prepare for possible biological differences should we ever come into contact with extraterrestrial life. By being aware that the diversity of life in the universe far exceeds our imagination, we can react more openly and respectfully to these potential encounters.

Exploring the biological properties of extraterrestrial life would not only expand our scientific boundaries but also revolutionize our imagination and understanding of life itself. It would help us question our own assumptions and prejudices and develop new perspectives on the existence and purpose of life in the universe. As a guide for potential interaction with extraterrestrial life, we should be open, attentive, and willing to learn. We may find that their biological properties and way of life are remarkably different from what we are accustomed to. These differences should be considered as enrichment that can help us better understand the complexity and beauty of the universe.

Ultimately, the exploration of the biological properties of extraterrestrial life is an exciting but also a journey fraught with many unknowns. It is a path of discovery, wonder, and awe at the immense diversity of the universe. As we continue to seek answers, we should be aware that we may only be scratching the surface of what is possible, and that the true secret of extraterrestrial life is still waiting to be discovered.

Psychological characteristics

In the infinite vastness of the universe, where we represent only a tiny speck, the possibility of encountering extraterrestrial life looms. But how might these alien beings think and act? Understanding their psychological properties presents us with a significant challenge because our previous investigations are limited to the behavior of humans and animals here on Earth.

To prepare for possible psychological properties of extraterrestrial life forms, we must expand our knowledge of perceptions, learning, and emotional states. It makes sense to explore information processing, storage, and retrieval in alien beings. This can provide a better understanding of their cognitive abilities. How do they perceive their environment? How do they think and solve problems? How does their memory function? But it's not just the individual aspects of psychology that are important. The social structures and interactions of extraterrestrial life forms should also be investigated. How do they shape their societies? What role does communication and cooperation play in their coexistence? By examining social behaviors, we can develop a better understanding of their psychological properties and prepare ourselves for interaction with alien cultures.

Another important aspect is the question of moral and ethical concepts. What is right or wrong for us may not necessarily apply to extraterrestrial life. Therefore, it is crucial to question our own moral concepts and ethical standards and openly discuss possible differences. Only in this way can we create a foundation for respectfully interacting with extraterrestrial life.

The emotions of extraterrestrial life forms also deserve our attention. Joy, fear, love, hate, curiosity, empathy, and aggression could be similar or entirely different in them. By exploring the emotional range, we can better prepare for different behaviors and reactions that we might encounter in a possible encounter.

It requires a broad spectrum of understanding and openness to grasp the diversity of thinking, feeling, and acting in other life forms. Only with a respectful and open-minded approach can we strive for harmonious interaction with extraterrestrial life, should we ever come into contact with them. The possibility of encountering the unknown opens up a world full of fascination and challenges that can expand our understanding of the universe and ourselves into unforeseen dimensions.

By preparing for physiological, biological, and psychological properties of extraterrestrial life, we take an important step in being prepared for possible encounters and utilizing the potential for interstellar exchange. It is significant that we set aside our own assumptions and prejudices and embrace the new and unexpected. It requires an openness of mind to engage with other ways of thinking, feeling, and acting that may seem foreign to us.

By being aware that our own experiences and perspectives represent only a small fraction of the entire diversity of life in the universe, we can create an atmosphere of respect and curiosity. The exploration of physiological, biological, and psychological properties of extraterrestrial life forms is not only of academic interest but also has practical implications for how we prepare for a possible encounter. It opens up the possibility of developing new technologies to facilitate communication and interaction with alien species. It can help us take appropriate measures to protect our own environment and the habitats of alien life. Moreover, it can also assist us in reflecting on and appreciating our own humanity and our place in the universe.

It is in our nature as a curious and exploratory species to seek answers to life's big questions. The search for extraterrestrial life and the understanding of its physiological, biological, and psychological properties are part of this fascinating adventure. It reminds us of how diverse life can be and how much we still have to discover. So, we can only look forward to the unknown that may await us in the depths of the universe. Exploring these new horizons can help us not only expand our understanding of the universe but also get to know ourselves better and appreciate our humanity. It is an extraordinary privilege to be part of this exciting age of exploration and to gradually understand the secrets of the universe.

As we continue our exploration into the realm of extraterrestrial psychology, it's crucial to acknowledge the complexities and uncertainties inherent in such endeavors. Our understanding of psychology is deeply rooted in the context of Earth and the evolutionary history of life on our planet. Extrapolating these principles to potential alien civilizations requires a delicate balance of imagination, scientific rigor, and humility.

One aspect to consider is the possibility of vastly different cognitive architectures among alien species. While humans rely predominantly on visual and auditory senses, extraterrestrial beings might possess entirely different sensory modalities or perceive reality in ways beyond our current comprehension. Understanding the neural substrates and information processing mechanisms underlying their cognition would be a monumental task, requiring interdisciplinary collaboration and innovative research methodologies.

Furthermore, the social dynamics of extraterrestrial societies are likely to be markedly different from our own. Concepts such as kinship, hierarchy, cooperation, and conflict may manifest in ways that challenge our preconceived notions. Exploring the cultural norms, communication systems, and value systems of alien civilizations could offer profound insights into the diversity of social organization across the cosmos.

Ethical considerations also loom large in our hypothetical encounters with extraterrestrial intelligence. As we grapple with questions of morality and interstellar diplomacy, we must confront the possibility of cultural relativism on a cosmic scale. What constitutes ethical behavior for one species may clash with the values of another, necessitating nuanced approaches to ethical negotiation and mutual understanding.

Moreover, the psychological impact of contact with extraterrestrial life cannot be overstated. The existential implications of discovering intelligent beings beyond Earth could provoke profound shifts in human consciousness and worldview. Psychologists and anthropologists would play a crucial role in studying the psychological responses of individuals and societies to such paradigm-shifting events, offering support and guidance in navigating the uncertainties of cosmic contact.

In essence, the exploration of extraterrestrial psychology represents a journey into the unknown, replete with intellectual challenges, philosophical conundrums, and transformative possibilities. By embracing curiosity, humility, and interdisciplinary collaboration, we can embark on this voyage with open minds and hearts, ready to confront whatever mysteries the cosmos may unveil.

Common communication foundations

To prepare for possible psychological characteristics of extraterrestrial life forms, we need to expand our knowledge of mental perception, learning, and emotional states. We need to explore the fundamentals of information processing, storage, but also the recovery of information from memory in order to better understand the intellectual capabilities of extraterrestrial beings. How do they perceive their surroundings? How do they think and solve problems? How does your memory work? Interstellar communication presents unique challenges as we attempt to contact extraterrestrial life forms. In order to enable an exchange of information at all, we should deal with the basic elements of communication and perhaps also question our own cultural ideas. We will look at various exchange options in order to create common ground.

Communication plays a central role in establishing connection and building mutual understanding between humans and extraterrestrial life forms. This is the only way we can exchange ideas and information in order to achieve common goals and avoid misunderstandings. We will recognize that communication does not only consist of verbal transmission, but must also include non-verbal elements such as body language and emotions. Language forms a key component for interstellar dialogue. But we should also note that different species may use different languages?? or even have completely different communication systems. Therefore, it is crucial to develop a fundamental understanding of the structure, meaning and diversity of language in order to enable effective communication with extraterrestrials.

Emotions, but also body language, play an important role in interstellar exchange. No matter what form of life it is, emotions are universal signals for certain states and can also represent reactions to certain courses of action. It is therefore important to take a closer look at the topic of non-verbal communication, i.e. body language. It requires openness, sensitivity and a willingness to acknowledge and respect differences to even allow communication with extraterrestrials. As a basis for interstellar communication, it is important to develop communication strategies that can reconcile the understanding of language, emotions, body language and cultural differences.

We would like to list some basic guidelines that can help us when communicating with extraterrestrial life forms:

Openness and curiosity: Let's be open to new experiences and ideas. Let's look at communication with extraterrestrials as a way of intercultural exchange and be curious about their ways of thinking and perspectives.

Empathy and sensitivity: Empathy allows us to put ourselves in the shoes of others and understand their feelings and needs. To create a deeper connection, we should strive to be empathetic and consider the perspective of extraterrestrial life forms. By noticing and respecting their emotions and reactions, we can increase trust and achieve more harmonious communication.

Clear and precise communication: Because language barriers and cultural differences can occur, it is important to convey clear and precise messages. Let's avoid complicated language or culturally specific expressions and use simple and universal concepts to communicate our intentions. Additionally, we can use visual or symbolic means to support our messages and minimize misunderstandings.

Patience and Respect: Interstellar communication requires patience as it takes time to overcome language barriers and develop a deeper understanding. Let's be respectful of the different communication styles and speeds of alien life forms. By giving them space to express themselves and waiting patiently for their reactions, we can create an atmosphere of trust. It can be assumed that our way of life and our communication represent a game of patience for extraterrestrial life forms. In this case, we can hope that we will also be shown respect and patience.

If we are able to adhere to these basic principles of communication and focus on empathy, sensitivity, clear messages, patience and a willingness to learn, we can create effective and enriching interstellar communication. It is our responsibility to build a bridge between different life forms and promote dialogue to achieve a deeper understanding of the universe and our own existence.

Preparation for extraterrestrial contact

Possible forms of contact

Humanity has always wondered what kinds of contacts with extraterrestrial life would be possible. This subchapter addresses this very question and highlights the different scenarios that could occur, as well as the opportunities and problems associated with them.

First, let's imagine peaceful contact. In a world where humans and extraterrestrial life forms interact peacefully, establishing successful communication is paramount. Possible language and cultural barriers must be overcome in order to develop a deep understanding of one another. It takes patience and openness to build trust and develop a positive relationship with the aliens.

But not every contact goes smoothly. There is also the possibility of conflict with an alien life form. In such moments it is crucial to act prudently and not react impulsively. Aggressive gestures or actions should be avoided to avoid escalation. Instead, it's important to remain calm and maintain an open line of communication. By closely observing the alien life form's behavior, its intentions can be interpreted and responded to appropriately. If necessary, non-verbal communication and gestures can be used to convey a message. However, violence should always be viewed as a last resort and only used to protect yourself or others.

In addition to conflict, there is also the possibility of a direct threat from an alien life form. In such situations, it is of utmost importance to act calmly and calmly so that the situation does not escalate further. Aggressive posture or weapons pointed at humans can pose a major challenge. To de-escalate the threat, speak calmly and slowly without making threatening gestures. Eye contact should be avoided as in some cultures (earth experiences) this can be interpreted as aggression. Instead, show a friendly demeanor and try to avoid eye contact. If possible, it is advisable to withdraw and leave the room to calm the situation. Additionally, local authorities or experts who have the necessary knowledge and technology to deal with such a threat should be contacted immediately.

We all want to help promote understanding of how to respond appropriately to various forms of contact with extraterrestrial life. It offers guidelines for dealing with peaceful encounters, conflict situations and threats. By knowing the correct course of action in every situation, we can improve our chances of successful interstellar communication while ensuring the safety and well-being of everyone involved.

Unfortunately, these are all speculative considerations, as we have no proven contact with extraterrestrial life. Nevertheless, it is important to address these questions and discuss possible scenarios in order to be prepared for the future. Only through a reflective approach can we overcome emerging challenges and further develop our interstellar communication skills.

Fears and prejudices

Some people probably want nothing more than contact with alien species. They are neither afraid nor insecure, but are full of expectations and joy. Of course, this could be a mistake, because we don't know how a possible life form will react to an encounter.

For other people, possible contact with unknown life forms would trigger fear and panic. This is about a healthy mix of emotions and respect for such an encounter.

We do not know what this encounter will look like, who will experience it and how it will end. But we should be aware that the first step is always the most important.

We should become aware of the fears and prejudices that exist within us. Consider your own feelings and reactions to the topic of extraterrestrial contact. Ask yourself if you have fears and if there are prejudices. If so, which ones, and how can these fears and prejudices influence your willingness to communicate? Everyone has fears about things they cannot assess. It can therefore be helpful to acquire knowledge and inform yourself in advance, as much as possible. Use information and educational sources to find out about scientific findings, but also previous research results, in libraries. When doing so, look for reputable sources to take a rational and informed standpoint.

Perhaps we should accept that this unknown may soon or perhaps already be part of our lives, that extraterrestrial life and extraterrestrial life forms could become a potential reality. Try to approach this unknown with curiosity and openness and put possible fears and prejudices in the background. Always remember that extraterrestrial contact offers opportunities to expand our understanding of life and our existence.

In order to reduce prejudices, it is helpful to empathize with the situation yourself. Try to put yourself in the alien's shoes. Look at the situation from their perspective and question how you would experience this encounter. Be open to all perspectives that come your way, while remembering that our human experience and knowledge are very limited. Alien life forms will certainly think completely differently and have different perspectives than we do. Two worlds or two universes meet here, and it could well be that the other side also feels something like uncertainty or fear. A joint solution to this prejudice is desirable.

Some people can combat their fears and prejudices through meditation. They can calm their minds and thereby better observe and understand their own emotions. This will help you reduce fears and adopt a more open attitude. Find like-minded people and get in touch with them to discuss the topic. Also discuss your fears and prejudices in order to find common solutions and relieve each other of this anxiety. Overcoming your own fears and prejudices is an ongoing process. The more you engage with information and basic knowledge, the sooner you will be able to conquer your fears or calm an uneasy feeling. It's completely normal to have certain fears and prejudices. However, you also have to work on overcoming these and questioning your own attitude.

In addition to the steps mentioned, various strategies can help you deal with your fears and prejudices:

Visualization: Imagine in your mind a positive encounter with extraterrestrial life forms. Visualize peaceful and respectful communication where both sides learn and benefit from each other. This exercise can help you put your fears in the background and develop positive expectations.

Education and research: Take advantage of the opportunities for training and research and engage with scientific findings and possible research results on the subject of extraterrestrial life. If you don't know something, knowledge can take away a lot of the fear.

Mindfulness: Practice mindfulness to become aware when fears or prejudices arise. Acknowledge these emotions without letting them overwhelm you and focus your attention on positive and constructive thoughts.

Openness to new experiences: Open yourself to new experiences and different perspectives. Take the opportunity to question your own beliefs and learn from others. This can help you reduce your fears and overcome prejudices.

Professional support: If you find that your fears and prejudices are severe and you are having difficulty dealing with them, do not hesitate to seek professional support. A therapist or counselor can help you understand your fears and develop strategies to manage them. (I don't want to unsettle you, but I openly admit that a therapist on this topic is a balancing act.) Here, exchanging ideas with like-minded people is more beneficial.

If you are willing to confront your own fears and prejudices and actively work to overcome them, you will create a solid basis for positive and open communication should you encounter aliens. Remember that every person is different and has their own specific life experience in terms of fears and prejudices. Be patient with yourself and hold that through your efforts you can become a more open and understanding communicator. It is only our responsibility to overcome our fears and prejudices and prepare for a possible encounter with extraterrestrial life. By treating each other with openness, curiosity and respect, we can seize opportunities to expand our understanding of life and existence.

Mental and emotional preparation

Mental and emotional preparation plays a crucial role when encountering extraterrestrial life forms. In order to optimally prepare for extraterrestrial contact, it is important to strengthen your own mental strength and resilience.

This chapter presents various techniques and exercises that can help you prepare mentally and emotionally for this extraordinary situation.

Imagine how exciting it will be to meet an alien civilization. Your thoughts are full of anticipation and curiosity. But at the same time, fears and insecurities also arise. What will the aliens be like? How will they communicate? In order to cope with such challenges, conscious mental and emotional preparation is crucial. Vividly visualize what an encounter with extraterrestrial life forms might be like. Imagine communicating, learning, and benefiting from each other in peace and respect. By imagining these positive scenarios, you can reduce fears and concerns and develop positive expectations.

The practice of meditation can help you develop a calm and clear mind. Regular meditation can improve your ability to concentrate, reduce stress, and help you be present in the moment. This is especially important when encountering alien life forms, as it allows you to fully concentrate on communication and interaction. A positive inner attitude is essential for extraterrestrial contact. Try to develop an open and respectful attitude and reduce possible prejudices or negative expectations. View extraterrestrial contact as a unique opportunity for learning and intercultural exchange. By cultivating a positive inner attitude, you can make better use of the opportunities that extraterrestrial contact offers. It is human to have insecurities and fears regarding extraterrestrial contact. It is important to become aware of these fears and actively work to overcome them. Identify your fears and try to figure out where they come from. Deal with them consciously and look for strategies to deal with them.

Don't forget to take care of your own mental and emotional health. Make sure you get enough sleep, a healthy diet and regular exercise. Find activities that bring you joy and help you reduce stress. By taking care of your own well-being, you will be able to better prepare for extraterrestrial contact and deal positively with the challenges that may arise. It can also be helpful to talk to other people who are also interested in extraterrestrial contact or are preparing for it. Look for forums, groups, or events where you can share your thoughts, questions, and experiences. Exchanging ideas with like-minded people can not only be supportive, but also offer new perspectives and insights.

Be prepared to reconsider and adapt your own ideas and beliefs. Extraterrestrial contact can bring new insights and information that can challenge our previous assumptions. By remaining flexible and open, you can better respond to unexpected situations and explore new possibilities.

Openness and respect

Extraterrestrial contact gives us the opportunity to come into contact with life forms that may have completely different cultural backgrounds and behaviors. In this chapter we will take a closer look at the importance of openness and respect for cultural diversity in the context of extraterrestrial contact. It's about how we can deal with the differences and which communication strategies and techniques help us to promote respectful and constructive collaboration.

It is important to understand that extraterrestrial life forms can potentially come from a variety of cultural backgrounds. Their values, norms, customs and beliefs will differ significantly from our own. It is therefore crucial not to use our own cultural ideas as a benchmark, but rather to be open to new perspectives. An open attitude is of great importance when we engage in extraterrestrial contact. We should be willing to rethink our own ideas and beliefs and to embrace the unknown, even if it contradicts our previous assumptions.

Empathy plays a central role in understanding cultural differences and dealing with them respectfully. By putting ourselves in the shoes of the extraterrestrial life forms and viewing the world from their perspective, we can learn to interpret their feelings, needs and actions in a cultural context.

To ensure effective communication with extraterrestrial life forms, intercultural communication strategies are essential. This includes active listening, using clear and concise messages, avoiding assumptions and prejudices, and being willing to ask questions and clarify ambiguities. Respect is a fundamental value when dealing with cultural diversity. Instead of dismissing the culture of alien life forms as inferior or strange, we should acknowledge their cultural traditions and practices and treat them with respect and dignity.

Despite cultural differences, there are often common foundations on which communication can be built. By looking for common interests, values? ?or goals, we can build bridges between different cultures and promote mutual understanding. Conflicts can arise when dealing with cultural diversity. These conflicts should be viewed in a constructive way. Solutions that are acceptable to both sides should be sought. A willingness to compromise and the search for win-win situations can help build harmonious relationships and enable long-term collaboration.

To enable open and respectful communication, it is important to develop an awareness of our own cultural assumptions, prejudices and stereotypes. By regularly reflecting on our thought patterns and ideas, this is the only way we can develop a conscious and sensitive attitude towards cultural diversity. This helps to reduce prejudices and promote open communication.

Openness and respect for cultural diversity are crucial to establishing successful extraterrestrial contact. By applying intercultural communication strategies, finding common ground, and being willing to adapt and be flexible, we can promote harmonious and productive collaboration with extraterrestrial life forms. It is up to us to see cultural diversity as an enrichment and to actively work to build bridges and enable respectful communication.

■

Communication strategies

Basic knowledge of diplomacy

Diplomacy refers to the art and science of relations between nations and other political entities.

Your goal is to resolve conflicts, reach compromises and reach agreements that take into account the interests of all parties involved. In the context of contact with extraterrestrial life forms, diplomacy plays a significant role as it involves establishing and maintaining a peaceful relationship.

If we want to establish a diplomatic relationship with an alien life form, we must first understand their cultural differences and values. The cultural differences between us and an alien life form could be enormous. The life form could think and act completely differently than we do, potentially leading to misunderstandings or even conflict.

Therefore, we must be aware that our own cultural values?? may not be universally applicable and that we must strive to understand and respect the other party's cultural differences. Diplomacy in communication is an important aspect. The way we express ourselves can be very different and vary greatly from one life form to another. Therefore, it is important to understand and, if necessary, learn the other party's language and communication methods to ensure clear and effective communication.

Another important factor in diplomacy is the ability to compromise. When negotiating with an alien life form, it may be necessary to make concessions in order to establish and maintain a connection. It is important to keep your own interests in mind, but also consider the needs and interests of the other party.

Finally, it is important to be patient when dealing with an alien life form. Establishing a diplomatic relationship can take a lot of time and patience, especially when there are major cultural differences. It is important to build long-term relationships based on respect and trust.

Diplomacy can be a crucial factor when dealing with extraterrestrial life forms. We must strive to understand and respect their cultural differences, learn effective communication methods, compromise and be patient in order to build long-term relationships. This is the only way we can establish and develop a peaceful and successful connection to extraterrestrial life.

It is also important to ensure that only people who are genuinely interested in peace and not in military conflicts conduct negotiations at the diplomatic level. It is essential to identify individuals at the highest level who not only have diplomatic skills, but are able to conduct diplomatic negotiations with truly peaceful intentions. Unfortunately, the world is characterized by numerous conflicts in which innocent people lose their lives. It is unfortunate that power-hungry individuals sometimes start wars that could actually be avoided. We can only hope that we will be able to send the right people to these negotiations. People who are committed to peace and are not driven by personal power interests.

Challenges

challenges

Special considerations and challenges that may arise when developing diplomatic relations with extraterrestrial life forms are addressed here. Diplomacy is an essential part of any successful intercultural relationship.

Success in establishing and maintaining positive relationships with extraterrestrial life forms depends on diplomatic negotiation skills. It is crucial to understand possible cultural differences and barriers that can hinder successful communication. Alien life forms will have completely different concepts of time, space, community and power compared to us humans. Therefore, it is important to understand these differences while still creating a foundation of respect and understanding in relationship building.

The challenge is to find a common language as a basis for communication with extraterrestrial life forms. The language that visitors from space will have will be fundamentally different from the language that we humans speak. What is needed here are people who are able to act as interpreters or translators in order to overcome barriers in communication and lay a basis for further interactions.

One possible approach is to involve experts in linguistics and intercultural communication to not only bridge linguistic differences but also understand cultural nuances. Precise and culturally sensitive communication is crucial to avoid misunderstandings and create a positive basis for the relationship. Therefore, we should strive to involve language experts with a deep understanding of the cultural backgrounds of extraterrestrial life forms.

There is currently a lack of clarity among the public about how communication could be achieved with possible visitors from space. Why this is so remains unknown. There is confusion about why sightings are concealed or concealed, and why objects that are clearly filmed or photographed are repeatedly dismissed as weather balloons. The fact that governments are keeping everything related to this issue under wraps also raises questions and contributes to uncertainty. Transparent communication about the possibility of an encounter as well as the opportunities and risks would be important in order to build trust and reduce uncertainty. However, there is currently a lack of clear information about how potential communication with extraterrestrial life forms could be designed.

In any case, a successful diplomatic relationship requires an understanding of cultural differences, careful consideration of language barriers and, of course, preparing people for possible contact with extraterrestrial life forms. Because when extraterrestrial creatures visit Earth, it will be humans who come into contact with them. Regardless of identity, we should prepare ourselves mentally.

Goals and interests

When considering contact with extraterrestrial life forms, it should be borne in mind that their goals and interests may differ significantly from our own. These differences may arise from the physiological, biological and cultural characteristics of the extraterrestrial life forms. It is conceivable that our planet's resources are of great importance for extraterrestrial life forms. Their visit could indicate that they may want to exploit the components of our planet that are valuable to them. Such goal differences could lead to conflict if they do not align with our own goals or if we are unable to meet their needs and demands.

In addition, studying our planet and our species for extraterrestrial life forms could also be of interest, be it for scientific, cultural or even military reasons. In such a case, it is important to establish diplomatic relations to ensure that their research and interactions are consistent with ethical and moral principles.

Regardless of the motivations behind visitors coming to our Earth, we should hope that extraterrestrial life forms have peaceful intentions and are interested in cooperating with us. In this case, it would be of great importance to identify our common interests and work together to build a positive relationship. However, in a worst-case scenario, there is also the possibility that extraterrestrial life forms could pose a threat to us, be it through aggressive actions or through the transmission of diseases or viruses. In such a situation, we would need to take appropriate precautionary measures to ensure our safety.

Goals and interests of extraterrestrial life forms will not be fixed or constant. Your goals and interests may be variable and may evolve over time or adapt to new circumstances.

Therefore, it is necessary to be flexible and understand their goals and interests in order to respond appropriately.

To understand the goals and interests of extraterrestrial life forms, it is important to be open to dialogue and get to know each other. By sharing information and finding common interfaces, we can develop a better understanding of their motivations and intentions and overcome possible misunderstandings and biases. Fears and reservations about the unknown are often based on a lack of information and assumptions. Open and respectful dialogue can help break down these barriers and build trust.

Creating a framework for knowledge sharing and collaboration at different levels can help identify common interests and promote a positive relationship with extraterrestrial life forms. This can include the exchange of scientific knowledge, technological innovations or cultural experiences. It is definitely worth exploring and understanding the goals and interests of alien life forms in order to establish a successful diplomatic relationship. Openness to dialogue and the willingness to find common interfaces are of great importance. Through the transfer of knowledge and collaboration, we can create a positive and enriching relationship with extraterrestrial life forms. At the same time, we should always remember that contact with extraterrestrial life forms is a complex process that requires time, patience and a careful approach.

Consideration of ethical principles and values? ?is of paramount importance when exploring the goals and interests of extraterrestrial life forms. When working with them, it is important to ensure that we respect their rights and interests and avoid exploitation or abuse. The ethical aspect plays a central role when it comes to ensuring responsible and respectful interaction with extraterrestrial life forms. It is crucial that we respect their needs and rights and act on a basis of respect.

It is also possible that some extraterrestrial life forms have more advanced technology or knowledge than we do. In such cases, we should be open to sharing knowledge and technologies to benefit from their progress. At the same time, we must ensure that these exchanges are reciprocal and that we protect our own interests.

Another challenge could be that extraterrestrial life forms may have a completely different perception of space and time than humans. This can make it difficult to communicate and understand their goals and interests. It requires flexibility and openness to adapt to new ways of thinking and develop alternative approaches to interact with them effectively.

In addition to the goals and interests of the extraterrestrial life forms, we should also consider our own goals and interests. We must clearly define our priorities and ensure that collaboration with extraterrestrial life forms is consistent with our values? ?and goals. A careful consideration of the potential advantages and disadvantages as well as long-term strategic planning are required.

But we also need to be sure that we are taking a holistic and responsible approach to researching the goals and interests of extraterrestrial life forms. This requires close collaboration between scientists, diplomats, governments and the international community to develop policies and protocols to ensure sustainable and respectful interaction with extraterrestrial life forms. The process of exploring and building relationships with extraterrestrial life forms is a fascinating and complex undertaking. It requires openness, curiosity and the willingness to go beyond our own limits. If we act with respect, knowledge and caution, we can develop a positive and enriching relationship with extraterrestrial life forms that can help us explore the mysteries of the universe and expand our understanding of life.

The certainty is that we should be aware of the challenges that contact with extraterrestrial life forms can bring. Through openness, willingness to dialogue, exchange of knowledge and consideration of ethical principles, we can respond to your goals and interests in the best possible way and build a positive and sustainable relationship. This requires a collaborative approach at the international level to develop policies and protocols that ensure contact with extraterrestrial life is handled responsibly and respectfully.

Communication techniques and principles

We have already discussed important communication strategies for possible contact with extraterrestrial life forms, such as diplomacy, overcoming challenges, and considering goals and interests. However, it is equally important to master basic communication techniques as they play a crucial role in effectively communicating with potential aliens. One of these techniques is active listening, which shows understanding and attention to the living being. This includes fully engaging in the conversation, asking questions, and summarizing information. When possible contact with extraterrestrial life forms, it is of great importance to listen carefully and pay attention to both verbal and non-verbal signals in order to truly understand the other person.

Another important skill is clear and precise language in order to communicate effectively and express yourself clearly. Misunderstandings could be avoided by formulating thoughts and information clearly. It is advisable to use simple and understandable language and avoid technical jargon or complex terms. Especially in the case of possible extraterrestrial contact, where language barriers will arise, it is even more important to make the statements clear and simple. Asking questions is also very important. Questions are used to obtain information and advance the conversation. Through targeted questions, understanding can be deepened and new insights gained.

In the case of potential extraterrestrial contact, questions can be used to get to know the other person better, understand their perspectives and find common ground.

It is important to be open to questions and also encourage extraterrestrial life forms to ask questions in order to develop a comprehensive understanding.

Empathy also plays an important role. Empathy means putting yourself in the other person's shoes to understand their feelings, needs and perspectives. Potential alien contact may involve different cultural backgrounds and life experiences. Through sensitivity, a bridge of understanding can be built and respect for the other person's feelings and perspectives can be shown. Expressing interest in their culture, traditions and values?? and being willing to respect them is important.

Non-verbal communication such as gestures, facial expressions, posture and eye contact plays a universal role. It can help to overcome existing language barriers. It is important to pay attention to the other person's non-verbal signals and respond appropriately. Your own non-verbal signals also play an important role. By consciously training body language, non-verbal communication can be improved and misunderstandings can be avoided. However, it should be noted that gestures and posture can have different meanings in different cultures. Sensitivity and respect for cultural differences in non-verbal communication are therefore crucial.

Another important technique is adapting communication to the specific habits of alien life forms. This includes adjusting language, tone, speed and information density. The willingness to learn and use new communication tools and technologies to facilitate communication is of great importance, and flexibility and openness to meet the needs and preferences of the alien interlocutors are essential.

Furthermore, clarity and comprehensibility of the messages are crucial. It is important to avoid ambiguity and ensure that statements are clear and easy to interpret.

Using clear examples and illustrating ideas can help minimize misunderstandings. If necessary, visual aids such as pictures, diagrams or drawings can also be used to support statements.

Respectful communication is a key aspect of successful interaction in possible extraterrestrial contact. It is important to respect the opinions, beliefs and cultural differences of the interviewees. Prejudices or derogatory statements should be avoided. By listening carefully and acknowledging the perspectives and contributions of the other person, trust can be built and a positive conversation atmosphere can be created.

These basic communication techniques and principles are of great importance to ensure effective and successful communication during possible extraterrestrial contact. By strengthening communication skills and preparing for possible contact with extraterrestrials, a positive relationship and thus a deeper understanding can be built. Be open, curious and respectful of extraterrestrial life forms and use communication as a bridge for meaningful exchange of information, ideas and experiences.

Non-verbal communication, body language

We believe that most communication with extraterrestrials could occur through non-verbal means. Given their ability to visit our planet with UFOs, we must admit that they are spiritually advanced beings. Since we have not yet visited other planets, we must also admit that these creatures are more intelligent than us. We will most likely not speak the same language, but it still makes sense to find a common denominator to enable communication. This common denominator will largely take the form of nonverbal communication. Therefore, we would like to pay more attention to this area.

Nonverbal communication and body language

Nonverbal communication and body language play a crucial role in extraterrestrial contact. While language and words are important means of communication, non-verbal signals often convey additional information and contribute to deeper understanding. In this chapter, we will explore the importance of nonverbal communication in an alien contact and provide practical tips on how to improve your body language skills.

gestures and facial expressions

Gestures and facial expressions are important elements of nonverbal communication because they can express emotions, intentions and messages without words. In extraterrestrial contact, gestures and facial expressions can vary, but there are some universal signs that can indicate communication. Carefully observe the gestures and facial expressions of your alien interlocutors and try to interpret their meaning. Also pay attention to your own gestures and facial expressions to support your statements and convey clarity.

Posture and posture

Posture can say a lot about your attitude and how you feel. An upright posture signals openness and confidence, while slumped shoulders or a defensive posture can express insecurity or rejection. In extraterrestrial contact, it is important to adopt an open and approachable posture in order to signal trust and sympathy. Try to maintain an upright posture and avoid closed positions, such as crossed arms or downward gaze.

Eye contact

Eye contact plays a crucial role in communication. It can convey intimacy, attention and trust. However, in extraterrestrial contact there may be differences in the meaning and handling of eye contact. Some alien life forms may have different viewing habits or consider direct eye contact inappropriate. Be attentive and pay attention to the reactions of your interlocutors in order to respond appropriately to their cultural differences.

Cultural differences in body language

It is important to realize that body language can be heavily influenced by cultural influences. What is considered friendly and welcoming in one culture may be perceived as rude or even aggressive in another. These differences become particularly relevant in the context of extraterrestrial contact. We would like to emphasize: Educate yourself about the cultural norms and habits of the alien life forms you wish to communicate with. But, as we all know, there are no fixed norms in this context. Therefore, it is crucial to respect their cultural differences and adjust your own body language accordingly.

Sensitivity to non-verbal signals

It is important to be sensitive to non-verbal signals, both from the extraterrestrial life forms and from yourself. Non-verbal signals can be subtle and convey information about emotions, approval or disapproval. Pay close attention to the posture, gestures, facial expressions and other non-verbal expressions of your conversation partners. Try to interpret the meaning of these signals to develop a more complete understanding of their messages.

Adapting to different communication styles

Alien life forms can have different communication styles, which are also reflected in their non-verbal communication. Some may be direct while others prefer indirect or symbolic forms of communication. Be flexible and adapt to the communication style of your interlocutors to ensure a smooth interaction.

Body language as a complement to language

In extraterrestrial contact it may happen that the language is not sufficiently available or communication problems arise. In such cases, body language becomes even more important. Use your non-verbal skills to support and convey your messages. Be sure to consciously use your gestures, facial expressions and posture to clearly express your intentions.

Mindfulness and empathy

Remain mindful and attentive to the non-verbal signals of your conversation partners. Show empathy and try to put yourself in their shoes. This will help you notice subtle cues and emotional moods that may not be expressed verbally. By responding empathetically and respectfully, you can promote trusting and harmonious communication.

Physical distance

Physical distance varies depending on cultural influence and can also be interpreted differently in extraterrestrial contact. Some alien life forms can maintain a greater distance, while others prefer a certain proximity. Respect the personal boundaries of those you are speaking to and adjust your distance accordingly. However, be careful not to appear too close or too distant to create a comfortable communication environment.

Authenticity and congruence

Be authentic in your nonverbal communication and ensure congruence between your verbal and nonverbal messages. Conflicting body language can lead to misunderstandings and affect trust. Make sure your body language supports your true intentions and feelings and that you are authentic in your overall communication.

Continuous learning and improvement

By developing your non-verbal communication skills and consciously incorporating them into extraterrestrial contact, you can increase the effectiveness of your communication and create a deeper connection. Be patient, open to learning, and use every interaction as an opportunity to improve your skills.

Conclusion

Nonverbal communication and body language play a crucial role in extraterrestrial contact. While language conveys important information, non-verbal signals contribute to understanding on a deeper level. By increasing your awareness of gestures, facial expressions, posture, eye contact, and other nonverbal signals, you can improve your ability to communicate with extraterrestrial life forms. It is important to be aware of cultural differences and adapt to avoid misunderstandings. Through mindfulness, empathy and respect, you can promote harmonious communication and build a trusting relationship. Authenticity and congruence between verbal and nonverbal messages are also crucial.

Remember that nonverbal communication is an ongoing learning journey. Be open to feedback, observe your conversation partners carefully and consciously use your nonverbal skills to enable successful communication in extraterrestrial contact.

Exchange options

Easy exchange opportunities and understanding of different ways of thinking!

During extraterrestrial contact we encounter different ways of thinking, perspectives and cultural backgrounds. To enable effective communication, it is crucial to create easy exchanges and develop an understanding of this diversity. This chapter presents strategies to facilitate communication with extraterrestrial life forms and promote mutual understanding of different ways of thinking.

To understand the mindset and perspective of an alien life form, it is important to broaden our own perspective. Take the time to reflect on your own way of thinking and embrace new perspectives. Ask questions and listen actively to understand other people's motivations and beliefs. By changing your perspective, you can break down barriers and create a deeper connection. Even if there are big differences between cultures, it is important to find similarities in order to create a basis for exchange. Focus on shared interests, values, or goals. Search for or find topics that are universal and provide a space for open discussion. By finding common ground, you can build bridges and increase trust.

In intercultural communication, it is essential to cultivate respect for different viewpoints. Accept that other cultures have different values?? and beliefs that shape their perspective. Practice tolerance and openness to opinions that differ from your own. Avoid hasty judgments and show respect for diversity of thought. Cultural differences can lead to misunderstandings if they are not recognized and taken into account. Find out about the cultural backgrounds and customs of the extraterrestrial life forms you want to communicate with. Pay attention to non-verbal signals, taboos and cultural norms. Show sensitivity and adapt your communication accordingly to promote understanding and respect, because to minimize misunderstandings, it is important to communicate clearly and concisely. Use simple and clear language to convey your messages. Avoid cultural idioms that may not be understandable to the alien life form. Listen for mutual feedback and make sure you are properly understood. Active listening is a crucial aspect of promoting understanding of different ways of thinking. Concentrate on being truly present and fully following the person you are speaking to. Give him your full attention, do not interrupt and ask questions to clarify misunderstandings. By actively listening, you show interest and respect for the other person's perspective.

Be open to new information and learning experiences. In extraterrestrial contact, you may be confronted with entirely new ways of thinking that challenge your own assumptions and beliefs. Be willing to rethink your own biases and preconceptions and embrace new perspectives. By maintaining your curiosity and keeping an open mind, you can develop your communication skills and gain a deeper understanding of other ways of thinking. Sometimes words alone may not be enough to convey complex ideas or emotions. Therefore, also use other creative forms of communication such as drawings, symbols or physical expressions. Visualizations can help make abstract concepts tangible to create a deeper connection. Be willing to explore and embrace alternative forms of expression.

Communicating with extraterrestrial life forms can be challenging and often requires patience and resilience. Don't give up when difficulties arise or misunderstandings arise. Be patient and view communication as a process that requires time and adjustment. Remain open to learning from mistakes and see challenges as opportunities for personal growth.

Building trust is an important aspect of communicating with extraterrestrial life forms. Show yourself to be a reliable conversation partner who is open, honest and respectful. Fulfill promises and be willing to take responsibility for your statements and actions. By building trust, you create a solid foundation for effective and respectful communication.

By using simple exchanges and developing an understanding of different ways of thinking, you can improve communication in extraterrestrial contact and create a deeper connection. Be open, flexible and respectful of other perspectives and use every interaction as an opportunity to grow and expand your own thinking.

Practical applications

In our guide we would like to guide you through practical applications that will help you improve your communication skills and develop increased intercultural sensitivity.

The chapter on practical application seems logical and relevant to everyone in every situation. Even among people on Earth, there are always opportunities to improve communication skills. It is enough to consider how we communicate with other people – be it through verbal expressions or non-verbal techniques.

Each of us communicates through facial expressions and decoding signals.

When communicating with extraterrestrial life forms, it could be equally important to use their abilities to perceive and interpret communication signals. Imagine how fascinating it would be to develop a universal language that transcends the boundaries of language and culture. This requires not only a deeper understanding of their possible signals, but also the ability to recognize subtle nuances and intricacies in communication.

Through conscious practice in everyday life, whether observing animals or in interpersonal interactions, you can sharpen your senses and learn to react attentively to hidden messages. Remember that it is not always words that can convey the full range of thoughts and emotions.

A crucial aspect is developing your own empathy. By trying to put yourself in other people's shoes, you can not only avoid misunderstandings, but also develop a deeper connection with those around you and potentially with extraterrestrial life forms.

The idea is to create a space for a deeper understanding of diversity in communication while emphasizing the universal aspects that connect us all as intelligent life forms. As we move forward, we will look at practical exercises and techniques to make these concepts applicable in your everyday life to take your communication skills to a new level.

In this way, you not only learn verbal communication, but also the importance of body language and facial expressions, which are of immense value in interpersonal relationships. Everyone should strive to sharpen their skills in this area because what one person perceives is not always what the other person wants to express. What is important here is the perception and interpretation of communication signals, and these skills should be strengthened. Regardless of which living being you make contact with, it is always communication that connects living beings with one another. There are even groups dedicated to communication strategies for dealing with misunderstandings.

By simulating different scenarios, you develop a deeper understanding of different ways of thinking and learn to deal effectively with possible challenges.

If you want to deepen your knowledge in this area, keeping a journal could be a useful exercise. This journal is designed to reflect on your own communication strategies and over time you will recognize progress, challenges and experiences. I am convinced that you already have these skills, but there are also people who have not yet dealt intensively with this part of communication. With simple tools, communication skills can be strengthened and prepared for a wide variety of communication situations.

We would like to convey that a deeper understanding and correct interpretation of communication is not only useful in contact with extraterrestrial life forms, but can also overcome or even prevent many problems in everyday life. The ability to interpret non-verbal signals could not only clarify misunderstandings, but also help to strengthen interpersonal relationships and avoid potential major conflicts, whether in the form of war or divorce. The power of communication lies not only in words, but also in the art of understanding the unspoken messages of the other person.

Existing language for aliens

Lincos is a special form of interstellar communication developed as a linguistic project to enable possible communication with extraterrestrial life. The term "Lincos" is derived from "lingua cosmica", which can be translated in German as "cosmic language".

The concept was developed by mathematician and scientist Hans Freudenthal in the 1960s.

The basic idea behind Lincos is to create a universal language based on logical and mathematical principles. The assumption is that mathematical concepts and logic could serve as a basis for interspecific communication because they are considered universally understandable.

The Lincos Project posits that advanced alien civilizations capable of interstellar communications may share an understanding of mathematics and logic.

An important aspect of Lincos is that it is intended to enable not only the transmission of information, but also the communication of concepts and abstract thinking. The project involves the development of a language that not only allows the representation of numbers and facts, but also the transfer of basic mathematical principles, as well as science and ethics. Lincos relies on the use of mathematical symbols and concepts that are considered universally understandable. These include, for example, the representation of natural numbers, geometric figures and basic mathematical operations. The idea is that advanced alien civilizations capable of interstellar communications will be able to recognize and interpret these basic mathematical concepts.

An interesting aspect of Lincos is the integration of ethical principles.

The concept takes into account that successful communication should not only include sharing information, but also conveying ethical values. This reflects the assumption that an advanced extraterrestrial civilization communicating with us may also have an interest in the moral and ethical aspects of our society.

However, the Lincos project remains speculative and hypothetical as we have not yet received any detectable signals from extraterrestrial life. Nevertheless, it represents a fascinating attempt to grapple with the challenge of interstellar communication and provide a theoretical basis for possible exchanges with extraterrestrial life.

∎

Challenges and solutions

Different scenarios

In this section, we discuss various scenarios that could occur once we encounter alien life forms. We cover topics such as potential collaboration, cultural and scientific exchange, as well as possible challenges and threats from extraterrestrial life forms. By examining these scenarios, we would like to prepare for a possible future with extraterrestrial life and derive appropriate recommendations for action.

Our goal is to provide guidance on what steps should be taken to prepare humanity for possible impacts of contact with extraterrestrial life. In doing so, we take into account ethical and moral considerations, such as the protection of lives and resources. We strive to provide recommendations for preparing for possible threats from extraterrestrial life forms.

In addition, we are preparing ideas for how humanity can shape its relationships with extraterrestrial life in the future. We highlight potential benefits of collaboration, such as expanding our knowledge of the universe and developing new technologies. At the same time, we discuss potential challenges and risks in cooperation with extraterrestrial life forms.

Overall, this chapter provides important information and recommendations for action to prepare us and humanity for a possible future regarding interactions and encounters with extraterrestrial life.

Effects of possible contact

The chapter "Impacts of possible contact" delves deeply into humanity's future in relation to extraterrestrial life and focuses on preparing for the potential impacts of such contact. It is of great importance that we prepare for this situation in order to be able to respond appropriately.

An essential measure is to fully inform the public about the possible effects of contact with extraterrestrial life and to prepare them for this possibility. Humanity should develop awareness of the issue and be aware of the potential occurrence of such an event in order to respond appropriately. Providing information can help reduce fear and uncertainty and promote acceptance of a future encounter with extraterrestrial life.

Additionally, it is advisable to develop protocols and procedures that can be followed in the event of contact with extraterrestrial life. These protocols should cover various aspects, from contact to diplomatic relations and possible cooperation with extraterrestrial life forms. It is important to establish clear guidelines to ensure orderly and respectful treatment of these extraterrestrial life forms while protecting our own interests.

In order to respond appropriately to possible contact with extraterrestrial life, it is of great importance that scientists and government agencies advance research and development in relevant areas such as space technology, communications and linguistics. These advances enable improved interaction and communication with extraterrestrial life forms and support the emergence of in-depth knowledge of their characteristics, needs and intentions.

Successful handling of contact with extraterrestrial life requires close international cooperation. It is essential that countries share resources and knowledge and develop a common approach to the issue. The exchange of information, expertise and experiences enables us to adopt a coordinated and cooperative stance and to create a fundamental basis for intercultural understanding and cooperation with extraterrestrial life.

Contact with extraterrestrial life raises numerous questions that must be clarified in advance. We must address these questions and develop clear principles and guidelines to ensure that we are able to address ethical challenges and deal responsibly with extraterrestrial life. Protecting life, maintaining integrity and respecting other intelligent species should be fundamental principles of our actions.

Preparing for possible contact with extraterrestrial life requires a wide range of measures. This includes comprehensively informing the public about the possible effects and opportunities of such contact. The development of protocols and procedures ensures that we can interact with extraterrestrial life forms in a structured and respectful manner. Promoting research and development in relevant areas allows us to develop technological and communication skills necessary for effective interaction with extraterrestrial life. Close international collaboration is crucial to ensure a coordinated and collaborative approach. Finally, clarifying ethical and moral issues is of great importance to ensure that we deal with extraterrestrial life in a responsible and ethical manner.

By taking these measures, we can prepare for possible contact with extraterrestrial life and manage the potential impact in a positive way. It is critical that humanity proactively addresses this challenge and prepares for a future in which interaction and encounters with extraterrestrial life are possible.

Shaping relationships

Shaping relationships with extraterrestrial life requires a comprehensive consideration of various aspects. This includes promoting a diplomatic and cooperative attitude, because by building a positive foundation for relations, potential conflicts can be avoided and instead cooperation and mutual benefits can be promoted. This requires openness, understanding and a willingness to respond to the needs and perspectives of extraterrestrial life forms.

Another important factor in shaping relationships with extraterrestrial life is creating an open and transparent culture of dialogue and collaboration. Communication plays a crucial role in reducing misunderstandings and prejudices. By exchanging knowledge, experiences and cultural aspects, we can develop better mutual knowledge and a deeper understanding of each other. In parallel, it is important to prepare for possible effects of contact with extraterrestrial life. This requires developing plans and strategies to deal with potential threats. At the same time, however, we should also consider the opportunities and possibilities that could arise from collaboration. Cultural and scientific exchange with extraterrestrial life could lead to new insights, innovations and advances.

When developing relationships with extraterrestrial life, we must not forget that our actions can have an impact on these life forms. It is therefore important to act in an environmentally conscious manner and to take into account the protection of various habitats and resources. We should be aware of how our activities could affect the natural environment of extraterrestrial life forms and strive to find sustainable solutions.

Overall, developing relationships with extraterrestrial life requires comprehensive preparation and a holistic approach. It's about promoting openness, dialogue and collaboration, managing potential threats while recognizing the opportunities and benefits of working together. By placing these aspects at the center of our efforts, we can build positive relationships with extraterrestrial life and shape a future in which both humanity and extraterrestrial life forms benefit.

Benefits of collaboration

Cooperation with extraterrestrial life can offer numerous advantages. First, it could help deepen our understanding of the universe. Extraterrestrial life forms could give us insights into planets, stars and galaxies that would otherwise remain hidden from us. By sharing information, we could also expand our knowledge in various scientific disciplines such as physics, biology and chemistry.

Furthermore, collaboration with extraterrestrial life could lead to innovative technologies. An example of this would be the use of materials that are exclusively found in other parts of the universe and could be invaluable to our technology. Aliens may also have more advanced technologies that we could benefit from for our own good.

Cooperation with extraterrestrial life could also lead to cultural enrichment. By learning about their lifestyles, customs and arts, we could deepen our understanding of other cultures.

Ultimately, cooperation with extraterrestrial life could help solve some of humanity's greatest challenges. By joining forces with other intelligent life forms, we could collectively find solutions to global problems such as climate change, poverty and combating disease.

Still, it is important to consider the risks and challenges of collaborating with extraterrestrial life. We must ensure we are mindful of the potential impact on humanity and prepared for possible threats. Comprehensive preparation and planning will help us maximize the benefits of collaborating with extraterrestrial life while minimizing potential risks.

Possible challenges and risks

Although collaboration with extraterrestrial life can have many benefits, we must expect that there are challenges and risks that need to be considered.

If an extraterrestrial species lands on our planet, it stands to reason that their technical intelligence will be much more advanced than ours. Therefore, they may attempt to subjugate or possibly colonize our civilization. To be prepared for such a situation, we can only hope that humanity recognizes it in time and can prepare well to defend its own interests and freedoms.

Of course, there is also the possibility that our health may be threatened by unknown diseases or pathogens transmitted through contact with extraterrestrial life. We cannot avoid this risk, but we should try to minimize it by taking appropriate precautions and protective measures. Specialists from medicine and research are needed for this.

Visitors from space come to Earth with specific intentions. These could be resources that Earth has and that aliens are looking for. But the Earth as a living space could also be interesting for creatures from space. As human beings on earth, we also have to deal with this risk. In such a case, it would be important to have a robust defense capability to protect our interests and security.

Challenges could also lie in conflicts due to cultural differences. Diplomacy and respectful interaction with other cultures would be crucial in this case to ensure successful cooperation.

The possibility of alien life landing on our Earth will also have an impact on our religion and philosophy. Some religious and philosophical beliefs need to be questioned. This in turn could lead to conflicts and tensions. It is important to have an open and respectful dialogue on these issues to ensure peaceful coexistence.

All people should be aware that there will be some risks and challenges when coming into contact with extraterrestrial life. Careful planning, diplomacy and defense capability could be crucial to ensuring our security.

As we know, where there are risks, there are also challenges and opportunities.

Living beings from outer space offer opportunities and potential that we probably cannot yet imagine. If collaboration is possible, it could lead to significant scientific breakthroughs and technological advances. By sharing knowledge and resources, we could gain new insights into the universe and expand our understanding of physics, biology and other branches of science. We could also learn from extraterrestrial life forms how to better protect our environment and use existing resources more efficiently. This knowledge could improve our environmental practices. Collaboration could not only contribute to technological progress, but also help address global challenges such as climate change.

By getting to know different ways of life, traditions and values, we could expand our own worldview and develop a deeper understanding of the diversity of possible life in space. This could lead to a more respectful and tolerant coexistence not only with extraterrestrial life forms, but also within human society.

In order to fully exploit these challenges, risks, opportunities and potential, our international community is required to work together. A coordinated approach, based on common principles and guidelines, should form the basis for establishing uniform standards for dealing with extraterrestrial life and developing a global strategy for managing relations.

International organizations, such as the United Nations, would play an important role in this by providing a platform for dialogue and cooperation.

In order to be able to create a relationship with extraterrestrial life, a multidisciplinary approach is required. This is about minimizing risks, seizing opportunities and creating a sustainable and cooperative future. Openness, diplomacy, preparation and international cooperation can build positive relationships with extraterrestrial life and promise a promising shared future.

Misunderstandings and conflicts

When encountering extraterrestrial life forms, misunderstandings and conflicts will inevitably arise. Language barriers, different ways of thinking and diverse cultural backgrounds will inevitably lead to communication problems. Now it is up to us to use various strategies and techniques to identify such misunderstandings at an early stage and resolve conflicts to ensure effective and harmonious communication.

The first step in dealing with misunderstandings is to recognize that they can occur. Different languages?? and cultural backgrounds can lead to information being misinterpreted or distorted, resulting in the actual content not being understood. This awareness makes it possible to better confront and counteract misunderstandings that begin.

The key to resolving misunderstandings is active listening. Take the time to listen carefully to the person you are speaking to, regardless of where they come from. It's not just about superficial listening, but about understanding what the person actually means. Particular attention should be paid to verbal and non-verbal signals in order to understand the message behind the words. By asking targeted questions, you show interest and at the same time signal that you are trying to interpret the content correctly.

Minimizing misunderstandings is made easier when you communicate clearly and concisely. Therefore, avoid ambiguous statements and use simple and clear language that is useful. Adapt your communication to cultural circumstances and take possible language barriers into account. If necessary, supporting visual aids can be used to clarify your message.

Conflicts can arise when different opinions, needs and interests come together. In such cases, it is important to use conflict resolution techniques to find constructive solutions. Techniques such as finding common goals, negotiating compromises and creating a balanced solution can be helpful here. Be willing to compromise to acknowledge and understand the other person's point of view. Respect and compassion are essential elements for successfully dealing with misunderstandings and conflicts. Try to understand the other person's perspective and respect their opinion, even if you have a different opinion. Prejudice and hasty judgments should be avoided, and compassion for each other's needs can be beneficial.

In the event of a conflict between different parties, the use of mediation can be helpful. A neutral third person can help resolve these conflicts by creating a structure for dialogue and supporting the process. As a mediator between the parties, she can help find common solutions that are acceptable to everyone involved.

To resolve misunderstandings, it is important to be willing to compromise and be willing to learn. If you are flexible and consider your own views and positions, you can view the conflict as an opportunity for personal growth. In this way, you will improve your own understanding and expand your communication skills. A positive communication climate is crucial in order to resolve misunderstandings and conflicts more quickly. Promote open and respectful communication where all parties are free to express their opinions. Personal attacks should be avoided and instead of attacking, you should focus on solving the problem. Create a positive climate for communication to promote trust and enable better collaboration.

To resolve conflicts, it is important to continually work on developing your own communication skills. Take the time to reflect on and improve your communication strategies, be open to feedback, and take the opportunity to learn from others. By continually improving your communication skills, you can identify misunderstandings early and respond to them effectively.

By applying these strategies and techniques, you will be better prepared to resolve misunderstandings and manage conflict in extraterrestrial contact. Open, calm and respectful communication lays the foundation for successful cooperation and harmonious coexistence when dealing with extraterrestrial life forms.

Building trust and establishing cooperation

To ensure effective communication with extraterrestrial life forms, it is crucial to build trust and establish a solid foundation for collaboration. Various aspects of building trust and promoting cooperation can support this.

Open and respectful communication forms the basis for creating trust and promoting collaboration. Be sure to listen carefully to the aliens and treat their opinions and perspectives with respect. Avoid prejudice and be prepared to explain your own points of view without appearing dogmatic. An atmosphere of mutual respect creates the basis for trusting cooperation. Transparency and honesty are other important elements in building trust. Share relevant information about your intentions, motivations and goals. Avoid secrecy or withholding important information, as this could affect the aliens' trust. Openness and honesty help create an atmosphere of trust and collaboration.

Mutual understanding and sensitivity are crucial to establishing a connection with extraterrestrial life forms. Strive to understand and respect their culture, values? ?and perspectives. Be patient and open to other ideas and ways of thinking. By showing understanding, you can build bridges and find common ground to work together successfully. Building common goals and interests is a further step towards cooperation. Identify areas where you have common interests and where you can collaborate. This enables a collaborative approach and increases the likelihood of a successful collaboration. By finding common ground, you can create a foundation upon which trust can be built.

Consistency and reliability are also crucial to building trust. Keep your promises and show consistency in your behavior and actions. Avoid contradictory statements or unreliable behavior as this could affect the aliens' trust. Through reliability and consistency, you show that you are trustworthy and that you can be relied upon. If you have patience and perseverance, laying the foundation for a trusting cooperation with extraterrestrials requires time and commitment, just like with humans. Be patient, take the time to develop relationships, and don't give up when there are challenges. Continuous efforts and a long-term approach will help you build a solid foundation of trust and achieve successful collaborations with aliens.

Another important strategy for building trust and promoting cooperation is to address conflicts constructively. Conflicts are inevitable, especially when different cultures, values? ?and perspectives come together. It is important not to avoid or ignore conflicts, but rather to view them as an opportunity for growth and understanding. Commit to addressing conflicts openly and actively seeking solutions that are acceptable to everyone involved. This requires flexibility, a willingness to compromise and the attitude to explore alternative solutions. Active collaboration and shared knowledge exchange could be another approach to building trust and promoting cooperation. Share your knowledge and resources with the aliens and be open to acquiring new knowledge and skills from them. Learning and exploring together can increase trust and lead to closer collaboration. Also take advantage of the opportunity to develop joint projects and initiatives to promote meaningful collaboration.

Recognizing and valuing the contributions of extraterrestrials would represent another important aspect. Show interest in their skills, knowledge and culture. Recognize their contributions and show them respect and gratitude. Since the aliens have already found us and presumably have superior intelligence, we can learn from them. This helps strengthen mutual trust and promote positive collaboration.

It is desirable to maintain clear and open communication about expectations, boundaries and responsibilities. Clarify with the aliens what role each will play in the cooperation and what goals and obligations are associated with it. Through transparent communication, misunderstandings can be avoided and trust can be strengthened. In addition to the strategies mentioned, it is helpful to point out past successes and positive experiences. Point out milestones and achievements that have already been achieved to strengthen trust and raise awareness so that successful collaboration is possible. Building trust and establishing cooperation is an ongoing process that requires time and commitment. Be aware that trust does not happen overnight, but is built through ongoing positive experiences, openness and constructive efforts. By incorporating these approaches into your communication with extraterrestrials, you lay the foundation for successful collaboration and enable a fruitful exchange of knowledge and ideas.

Mental and emotional stability

Different techniques

Techniques for maintaining calm and clarity

In extraterrestrial contact, situations can arise that challenge our mental and emotional stability. In order to respond appropriately and constructively, it is important to know methods to maintain calm and gain clarity, or create. This subchapter presents various techniques to help you maintain your mental and emotional balance during extraterrestrial contact.

Many people find peace and clarity by meditating. I realize that this doesn't apply to everyone, but if you've never tried it, you can't know whether it might be helpful. So we will be aware of some techniques that can help in general.

Breathing is closely connected to our emotions and our mental state. Through conscious breathing we can calm down and focus our thoughts. A simple technique is abdominal breathing. Place one hand on your stomach and breathe deeply so that your stomach expands. Hold your breath for a moment and then exhale slowly. Repeat this process several times and feel your inner peace and clarity improve.

Meditation is an effective way to calm the mind and achieve inner clarity. Sit in a comfortable position, close your eyes, and focus on your breathing or a calming thought. Let thoughts pass by without getting carried away by them. Through regular meditation you can strengthen your mental stability and develop a calm and clear mindset.

Visualization techniques can help achieve a positive mental state to provide clarity. For example, imagine yourself thinking calmly, confidently, and clearly in a situation of extraterrestrial contact. Visualize yourself successfully communicating and establishing a harmonious connection with the extraterrestrial life forms. By creating such positive images in your mind, you can strengthen your mental and emotional stability.

Mindfulness means being conscious in the present moment and paying close attention to your thoughts, feelings and physical sensations. Through mindfulness exercises you can become aware of your own reactions and learn to consciously control them. Be aware of your emotions and thoughts without judging or giving in to them. This allows you to remain calm and clear in challenging situations. Taking care of your own well-being is always essential to maintaining mental and emotional stability. This applies to everyday life or even if you want to have peace and clarity in the area of?? an extraterrestrial experience.

Make time regularly for activities that bring you joy and relaxation. This can be exercise, reading, music, meeting friends, etc. Consciously create moments of peace and relaxation to recharge your batteries and strengthen your mental stability. Set clear boundaries and take time out whenever you feel the need. Also make sure you eat a healthy diet, get enough sleep and exercise, as these factors also have a big impact on your mental and emotional stability.

Regular reflection and self-reflection are important tools to promote your mental and emotional stability in any or possible extraterrestrial contact. Take time to reflect on your experiences and emotions. Ask yourself how you felt in certain situations and how you reacted to them. Identify possible triggers for stress or anxiety and consider how you can better deal with them in the future. Through this conscious self-reflection, you can better understand your own patterns and reactions and work specifically to influence them positively.

It is equally important to recognize early on that it can sometimes be helpful to seek and accept support from other people. Talk to people you trust, friends or family members, and of course also like-minded people, to talk about your expcriences and emotions in extraterrestrial contact. Often just sharing your thoughts and feelings can help you gain clarity and gain new perspective. In addition, professional support such as therapists or coaches can help you further strengthen your mental and emotional stability. That maintaining calm and clarity during extraterrestrial contact is crucial to communicating effectively and establishing a positive connection with the extraterrestrial life forms is a logical conclusion. The stress management techniques presented such as: relaxation, mindfulness, visualization and self-care will help you strengthen your mental and emotional stability. By regularly using these techniques, you can stay calm and act constructively even in challenging situations.

Insecurities and fears

When in contact with extraterrestrials, uncertainties and fears can arise as it is an unfamiliar and potentially frightening situation. It's important to acknowledge these emotions in order to find ways to deal with them. This section presents various strategies and techniques for managing insecurities and fears in extraterrestrial contact. Start by recognizing and accepting your own emotions and fears. Take time to understand the fears and insecurities you feel and how they affect your behavior and communication. By consciously dealing with your emotions, you can deal with them better and use them in a targeted manner.

Another way to deal with fears and insecurities is to talk to other people who are going through similar experiences. Sharing thoughts and feelings can help build a support network to find common solutions. Accepting that insecurities and fears are a natural part of extraterrestrial contact can help us adapt to the challenges. It's important to be patient with ourselves and understand that there are no quick fixes. Extraterrestrial contact is a unique experience in which we will gradually learn and grow.

Try to view uncertainties as opportunities for personal development. See extraterrestrial contact as an opportunity to grow beyond yourself and have new experiences. Accept that insecurities and fears are part of the process and that you can grow by facing the fears. There are various techniques for managing anxiety, as I have already described. It is important to have realistic expectations about extraterrestrial contact. Understand that it is an ongoing learning and growth experience and that not all insecurities and fears will disappear immediately. Set small goals and celebrate every progress you make. Maintaining a positive attitude and patience will help you deal with uncertainty.

Don't forget to take care of your own mental and emotional health. Make time regularly for self-care activities that bring you joy and reduce stress. These can be activities such as sports, meditation, reading, listening to music or exploring nature, but much more. By paying attention to your own well-being, you strengthen your mental stability in extraterrestrial contact.

Ultimately, it is important to accept the fact that certain insecurities and fears may always be present in extraterrestrial contact. It is a normal part of being human to feel uncertain in unfamiliar situations.

Learn to trust your own judgment and believe that you have the skills to deal with the challenges that extraterrestrial contact presents. By facing your insecurities and fears and using proven coping techniques, you can strengthen your mental and emotional stability during extraterrestrial contact. Keep in mind that everyone is individual and has their own approach to dealing with uncertainty. Find the strategies that suit you best and be patient with yourself during this exciting experience of extraterrestrial contact.

We should realize that there may not be a chance to refer to concrete backgrounds to reduce our insecurities and fears. Nevertheless, we can deal with our emotions, inform ourselves and exchange ideas with others in order to find a better way to deal with this unfamiliar situation.

Be patient with yourself to embark on the journey of extraterrestrial contact without any set ideas or handed down information. As we embark on this unique journey of experience, we have the opportunity to develop and grow ourselves. We can view our insecurities as opportunities for personal transformation. Every challenge and every encounter with the alien offers the opportunity to grow beyond ourselves and have new experiences. Recognizing and respecting our own boundaries can be very helpful in these situations. Each of us has individual comfort zones that we should identify. We should allow ourselves to communicate these boundaries and respect them. Setting boundaries allows us to feel safer and better manage our fears.

We should have realistic expectations of extraterrestrial contact. Must understand that this is a continuous learning and growth journey and that not all insecurities and fears will disappear immediately. With small goals that we set for ourselves, we can celebrate every progress we make. A positive attitude and patience will help us deal with the uncertainties that extraterrestrial contact brings.

Empathy and understanding

Yes, I am aware that you have already read these sentences, and it may happen that you will read them one or two times. Since everything is related in some way, this reference cannot be avoided. Remember that this creature comes from a completely different world than we are used to. In extraterrestrial contact, empathy and understanding are crucial. By striving to understand the perspectives and experiences of extraterrestrial life forms, we can create a deeper connection and minimize misunderstandings. Therefore, in this subchapter we discuss various techniques and strategies for promoting sensitivity and understanding in extraterrestrial contact.

A key aspect of empathy is active listening and perspective taking. By giving extraterrestrial lifeforms our full attention and seeking to understand their perspective, we can create a deeper connection. We should not only listen to their words, but also take into account their non-verbal communication and emotions. Additionally, we can try to put ourselves in their shoes and imagine how we would feel in similar situations. Asking open-ended questions is another effective method for promoting intuition and understanding in extraterrestrial contact. Instead of making assumptions, we should be curious and ask questions to develop a deeper understanding. Open questions encourage detailed answers and promote dialogue. It is important to remain respectful and not to ask overly personal questions that could violate the privacy of the extraterrestrial life forms.

To cultivate empathy, it can be helpful to consciously change your perspective. By asking ourselves how we would see the world if we were an alien life form, we can develop a deeper understanding. We look at their world through their eyes, taking into account their experiences, values?? and cultures. This allows us to connect on an emotional level. Reducing prejudices and clichés is another important aspect, as these can significantly impair communication and understanding in extraterrestrial contact. Be aware of what prejudices we might have in order to question them critically. When we recognize and reduce our own biases, we open ourselves up to new experiences and enable more open and empathetic communication. We should view the extraterrestrial life forms as individual beings and give them the opportunity to show their unique personalities and cultures.

In addition, it is important to develop cultural sensitivity. In extraterrestrial contact we may encounter different cultural backgrounds and value systems. To ensure smooth communication, it is important to demonstrate cultural sensitivity to respectfully deal with these differences.

Here are some strategies that can be used once contact occurs to promote cultural sensitivity:

Find out about different cultures: Learn about the diversity of alien cultures, their customs, traditions and value systems. Educate yourself about cultural norms and respectful behavior.

Be open and flexible: Be open to new perspectives and willing to rethink your own ideas and assumptions. Accept that there are different ways to understand and interpret the world.

Practice respectful communication: Make sure to practice respectful and compassionate communication. Avoid culturally inappropriate terms, gestures or topics. Be attentive to non-verbal signals and respect personal boundaries.

Sensitivity to language differences: Since it is assumed that there are language barriers, be patient and use alternative means of communication such as visual representations, symbols or sign language. Make an effort to overcome misunderstandings due to language differences.

Avoiding Clichés: Don't assume that all alien life forms from a particular culture or species have certain characteristics. Each individual way of life is unique, and stereotypes can lead to false assumptions and misunderstandings.

Dialogue and knowledge exchange: Promote dialogue and the exchange of knowledge and experiences. Through dialogue you can learn from each other and develop a deeper understanding of each other.

Empathy and understanding: Show empathy and understanding for the challenges and experiences of extraterrestrial life forms. Try to put yourself in their shoes and understand their perspective.

Fostering sensitivity and understanding in extraterrestrial contact allows us to connect on a deeper level and create common ground for effective communication and collaboration. By focusing on our empathic abilities and adopting a culturally sensitive attitude, we can achieve harmonious and respectful coexistence in extraterrestrial contact.

Insecurities and fears

When in contact with extraterrestrials, uncertainties and fears can arise as it is an unfamiliar and potentially frightening situation. It's important to acknowledge these emotions in order to find ways to deal with them. This section presents various strategies and techniques for managing insecurities and fears in extraterrestrial contact. Start by recognizing and accepting your own emotions and fears. Take time to understand the fears and insecurities you feel and how they affect your behavior and communication. By consciously dealing with your emotions, you can deal with them better and use them in a targeted manner.

Another way to deal with fears and insecurities is to talk to other people who are going through similar experiences. Sharing thoughts and feelings can help build a support network to find common solutions. Accepting that insecurities and fears are a natural part of extraterrestrial contact can help us adapt to the challenges. It's important to be patient with ourselves and understand that there are no quick fixes. Extraterrestrial contact is a unique experience in which we will gradually learn and grow.

Try to view uncertainties as opportunities for personal development. See extraterrestrial contact as an opportunity to grow beyond yourself and have new experiences. Accept that insecurities and fears are part of the process and that you can grow by facing the fears. There are various techniques for managing anxiety, as I have already described. It is important to have realistic expectations about extraterrestrial contact. Understand that it is an ongoing learning and growth experience and that not all insecurities and fears will disappear immediately. Set small goals and celebrate every progress you make. Maintaining a positive attitude and patience will help you deal with uncertainty.

Don't forget to take care of your own mental and emotional health. Make time regularly for self-care activities that bring you joy and reduce stress. These can be activities such as sports, meditation, reading, listening to music or exploring nature, but much more. By paying attention to your own well-being, you strengthen your mental stability in extraterrestrial contact.

Ultimately, it is important to accept the fact that certain insecurities and fears may always be present in extraterrestrial contact. It is a normal part of being human to feel uncertain in unfamiliar situations.

Learn to trust your own judgment and believe that you have the skills to deal with the challenges that extraterrestrial contact presents. By facing your insecurities and fears and using proven coping techniques, you can strengthen your mental and emotional stability during extraterrestrial contact. Keep in mind that everyone is individual and has their own approach to dealing with uncertainty. Find the strategies that suit you best and be patient with yourself during this exciting experience of extraterrestrial contact.

We should realize that there may not be a chance to refer to concrete backgrounds to reduce our insecurities and fears. Nevertheless, we can deal with our emotions, inform ourselves and exchange ideas with others in order to find a better way to deal with this unfamiliar situation.

Be patient with yourself to embark on the journey of extraterrestrial contact without any set ideas or handed down information. As we embark on this unique journey of experience, we have the opportunity to develop and grow ourselves. We can view our insecurities as opportunities for personal transformation. Every challenge and every encounter with the alien offers the opportunity to grow beyond ourselves and have new experiences. Recognizing and respecting our own boundaries can be very helpful in these situations. Each of us has individual comfort zones that we should identify. We should allow ourselves to communicate these boundaries and respect them. Setting boundaries allows us to feel safer and better manage our fears.

We should have realistic expectations of extraterrestrial contact. Must understand that this is a continuous learning and growth journey and that not all insecurities and fears will disappear immediately. With small goals that we set for ourselves, we can celebrate every progress we make. A positive attitude and patience will help us deal with the uncertainties that extraterrestrial contact brings.

Empathy and understanding

Yes, I am aware that you have already read these sentences, and it may happen that you will read them one or two times. Since everything is related in some way, this reference cannot be avoided. Remember that this creature comes from a completely different world than we are used to. In extraterrestrial contact, empathy and understanding are crucial. By striving to understand the perspectives and experiences of extraterrestrial life forms, we can create a deeper connection and minimize misunderstandings. Therefore, in this subchapter we discuss various techniques and strategies for promoting sensitivity and understanding in extraterrestrial contact.

A key aspect of empathy is active listening and perspective taking. By giving extraterrestrial lifeforms our full attention and seeking to understand their perspective, we can create a deeper connection. We should not only listen to their words, but also take into account their non-verbal communication and emotions. Additionally, we can try to put ourselves in their shoes and imagine how we would feel in similar situations. Asking open-ended questions is another effective method for promoting intuition and understanding in extraterrestrial contact. Instead of making assumptions, we should be curious and ask questions to develop a deeper understanding. Open questions encourage detailed answers and promote dialogue. It is important to remain respectful and not to ask overly personal questions that could violate the privacy of the extraterrestrial life forms.

To cultivate empathy, it can be helpful to consciously change your perspective. By asking ourselves how we would see the world if we were an alien life form, we can develop a deeper understanding. We look at their world through their eyes, taking into account their experiences, values?? and cultures. This allows us to connect on an emotional level. Reducing prejudices and clichés is another important aspect, as these can significantly impair communication and understanding in extraterrestrial contact. Be aware of what prejudices we might have in order to question them critically. When we recognize and reduce our own biases, we open ourselves up to new experiences and enable more open and empathetic communication. We should view the extraterrestrial life forms as individual beings and give them the opportunity to show their unique personalities and cultures.

In addition, it is important to develop cultural sensitivity. In extraterrestrial contact we may encounter different cultural backgrounds and value systems. To ensure smooth communication, it is important to demonstrate cultural sensitivity to respectfully deal with these differences.

Here are some strategies that can be used once contact occurs to promote cultural sensitivity:

Find out about different cultures: Learn about the diversity of alien cultures, their customs, traditions and value systems. Educate yourself about cultural norms and respectful behavior.

Be open and flexible: Be open to new perspectives and willing to rethink your own ideas and assumptions. Accept that there are different ways to understand and interpret the world.

Practice respectful communication: Make sure to practice respectful and compassionate communication. Avoid culturally inappropriate terms, gestures or topics. Be attentive to non-verbal signals and respect personal boundaries.

Sensitivity to language differences: Since it is assumed that there are language barriers, be patient and use alternative means of communication such as visual representations, symbols or sign language. Make an effort to overcome misunderstandings due to language differences.

Avoiding Clichés: Don't assume that all alien life forms from a particular culture or species have certain characteristics. Each individual way of life is unique, and stereotypes can lead to false assumptions and misunderstandings.

Dialogue and knowledge exchange: Promote dialogue and the exchange of knowledge and experiences. Through dialogue you can learn from each other and develop a deeper understanding of each other.

Empathy and understanding: Show empathy and understanding for the challenges and experiences of extraterrestrial life forms. Try to put yourself in their shoes and understand their perspective.

Fostering sensitivity and understanding in extraterrestrial contact allows us to connect on a deeper level and create common ground for effective communication and collaboration. By focusing on our empathic abilities and adopting a culturally sensitive attitude, we can achieve harmonious and respectful coexistence in extraterrestrial contact.

Look into the future

Developments

Technological developments and their effects on communication

 Possible extraterrestrial contact has always captured humanity's imagination. While we are currently in an era of intense research and discovery, it is exciting to look into the future and speculate on how communication with extraterrestrial life forms might evolve. Technological advances play a crucial role in this. In this subchapter we consider the potential impact of these technologies on communication with extraterrestrial contact.

Advances in Translation Technology:
 The language barrier has always been a challenge in extraterrestrial contact. But with rapid advances in translation technology, we may soon be able to translate languages?? in real time. Machine learning and artificial intelligence are already enabling impressive translation results. It is easily conceivable that future developments could even enable direct translation of ideas, where language no longer serves as a barrier but as a means of connection.

Telecommunication technology:
 Overcoming large distances is one of the greatest challenges in extraterrestrial contact. Future technologies could allow us to communicate faster and more efficient-

ly. Advances in space travel and quantum communications could dramatically reduce time periods and enable real-time communication. Long-distance communication devices could be developed to ensure seamless connection over long distances.

Holographic communication:
Imagine being able to interact with alien life forms in the form of lifelike holograms. Holographic communication could provide an immersive and realistic experience in which visual and auditory signals are transmitted in a virtual environment. Advances in holographic technology could allow us to consider nonverbal communication and body language in interstellar communication.

Virtual reality:
The use of virtual reality (VR) could also revolutionize communication in extraterrestrial contact. By moving into a virtual environment, we could create a common platform where we can meet and communicate with extraterrestrial life forms. VR offers the opportunity to convey complex concepts and information in a visually appealing way and create a deeper connection.

Neurological interfaces:
Another intriguing possibility is the development of neurological interfaces that allow thoughts and emotions to be shared directly. By connecting the brain to an interface, we could transmit our thoughts and emotions directly to

extraterrestrial life forms. This would enable a profound form of communication that transcends language and cultural barriers. By exchanging thoughts, we could develop a deeper understanding of each other and connect on a whole new level.

Artificial Intelligence and Machine Learning:

Artificial intelligence (AI) and machine learning will play an important role in future communication with extraterrestrial life forms. AI-based systems could be able to analyze and interpret complex information from various sources to help us communicate. They could help us better understand the non-verbal signals and cultural nuances of extraterrestrial life forms to avoid possible misunderstandings.

Ethics and responsibility:

With all of these technological advances, it is important to consider the ethical aspects of extraterrestrial contact. The responsible use of the developed technologies and respect for the privacy and autonomy of extraterrestrial life forms must be guaranteed. It is very important that we are aware of how these technologies are used and that we communicate in a way that maintains respect and integrity.

The future of extraterrestrial contact promises exciting developments in communications. Advances in translation technology, telecommunications, holographic communications, virtual reality, neurological interfaces, and artificial intelligence and machine learning are opening up new opportunities for interactions with extraterrestrial life forms. It is up to us to use these technologies responsibly and to establish open and respectful communication. By preparing for this future, we can fully realize the potential of extraterrestrial contact and welcome a new era of interstellar communication. The outlook for future technological developments in extraterrestrial contact is based on speculative ideas and possible scenarios. It is important to follow current developments in science and technology to stay informed about the latest findings. Technological preparations are also crucial to prepare for the possibility of extraterrestrial life. There are a number of technologies that can be developed to make dealing with extraterrestrial lifeforms easier.

One important technology that could be developed is communications technology. It will be important to find a way to communicate with extraterrestrial life forms, even if they have a completely different language or method of communication. Researchers could work to develop a universal communication technology capable of translating and interpreting all types of languages?? and communication methods.

Another important technology is space technology. It will be important for humanity to be able to travel quickly and safely into space to interact with extraterrestrial life forms. The development of fast, reliable and safe spacecraft will be crucial. Finally, medical technologies will also be of great importance in preparing for the possibility of extraterrestrial life. Humanity would need to be able to respond quickly to possible threats from viruses and diseases that could be transmitted by extraterrestrial life forms. The development of rapid and effective tests, as well as vaccines and cures, will be crucial.

It is also important that humanity prepares for all possible scenarios when it comes to contact with extraterrestrial life. Technological preparations are an important part of these measures and should be further developed by researchers and scientists around the world.

Successfully developing and deploying these technologies will require close collaboration between governments, scientists, technology companies and the international community. It will be important to share resources and information to find the best solutions and ensure that extraterrestrial contact on a global scale is consistent and coordinated and that important aspects such as security, privacy and cultural sensitivity are taken into account.

It is also of great importance that the public is actively involved in the discourse on extraterrestrial contact. This can be done through information campaigns, public forums, debates and educational institutions. By raising awareness and familiarizing people with current scientific knowledge, we can reduce prejudice and promote positive attitudes towards the alien species.

Finally, we should be aware that extraterrestrial contact will also have an impact on our own society and culture. It will raise new questions and force us to rethink our ideas about identity, spirituality, ethics and our place in the universe. It is important that we take an interest in these changes and actively participate in shaping a new era of interstellar communication.

Overall, the future of extraterrestrial contact is full of possibilities. Through proper preparation, use of appropriate technologies, ethical considerations and broad societal participation, we can use extraterrestrial contact as an opportunity to expand our understanding of the universe and usher in a new era of interstellar communication. But we must also be careful to continue pursuing scientific research and discoveries to stay current and expand our knowledge of extraterrestrial contact.

Significance for humanity

The significance for humanity of an encounter with extraterrestrial life goes far beyond the technical aspects and also concerns the philosophical and sociological implications that such an encounter could have on our civilization. In this section we will explore this topic and examine the potential significance for humanity of an encounter with extraterrestrial life. If this encounter were to occur, it would undoubtedly have a profound impact on our worldview and perspectives.

We would face an entirely new life form that would expand our ideas about existence and intelligence. The experience of an encounter with extraterrestrial life could make us rethink our position in the universe and redefine our identity in the cosmic context. It could lead to a paradigm shift, a fundamental rethinking of how we understand ourselves and our relationship to the world around us. The opportunity of this contact would undoubtedly lead to enormous progress in science and technology. The technologies being developed for communication and exchange with extraterrestrial life forms could also be applied to other areas. New insights into physics, biology and other branches of science could be gained by learning from extraterrestrial life forms. This could lead to breakthroughs in medical research, space technology and other areas.

Such an encounter would undoubtedly have an impact on our society and culture. It could lead to a new unity and cooperation on a global scale as humanity faces a common challenge and a common goal. Cultural diversity could be viewed as wealth and differences between humans and extraterrestrial life forms as enrichment. It could lead to a new appreciation of diversity and a global identity that transcends national borders. This contact could also lead to intense spiritual and philosophical reflection.

Encountering an extraterrestrial intelligence could raise questions about the meaning of life, our existence and our place in the universe. It could lead to new ways of thinking and spiritual approaches that challenge and expand our previous beliefs. Humanity could move to a new phase of consciousness development and gain a deeper understanding of the nature of the universe. An encounter with extraterrestrial life would undoubtedly bring challenges, but also opportunities. We would have to face the challenges of intercultural communication and find ways to overcome cultural differences and misunderstandings. It would be an opportunity to take our communication skills to a whole new level. In addition, new horizons would also open up, accompanied by an enormous increase in knowledge, both in the areas of science, technology, but also philosophy and spirituality. We could learn from their advanced high technology and expand our knowledge of physics, biology and other disciplines. Encountering extraterrestrial life could also help us address primary problems facing humanity, such as finding alternative energy sources, tackling environmental problems or improving our social systems.

A rendezvous with extraterrestrial life would represent a transformative experience for humanity. It would broaden our perspectives, advance our technology and shape our society and culture. It could lead to a time of change and growth, where we overcome our differences and build a new global community. The importance of an encounter with extraterrestrial life lies not only in the technical aspects, but above all in the possibility of better understanding ourselves and defining our place in the universe.

Approaches and projects

Research into extraterrestrial contact

In this chapter, various approaches and projects for further research and exploration of extraterrestrial contact are presented. An ever-growing number of researchers, scientists and organizations are committed to exploring the mystery of extraterrestrial life and finding ways to communicate.

1. SETI: In search of extraterrestrial intelligence

The Search for Extraterrestrial Intelligence (SETI) is a major research initiative focused on capturing signals from space that could indicate extraterrestrial intelligence. Large radio telescopes are used to listen to signals from space and search for patterns or targeted messages. Although conclusive evidence of extraterrestrial intelligence is still pending, SETI research has already produced promising results.

2. Space missions, robotic exploration and exoplanet research

Another approach to studying extraterrestrial life is through space missions and robotic exploration. Space probes and robots are sent to other celestial bodies to search for signs of life. Current missions, such as the search for liquid water on Mars or the exploration of the moons of Jupiter and Saturn, offer fascinating insights.

3. Exoplanet research: searching for habitable worlds

Exoplanet research has made significant progress in recent years. Astronomers can increasingly identify and analyze exoplanets using modern telescopes. The search for Earth-like exoplanets that could potentially have life-friendly conditions is an exciting area of?? this research.

4. Interdisciplinary collaboration and citizen participation

Encountering extraterrestrial life requires broad collaboration across different expertise and disciplines. Scientists, linguists, psychologists, philosophers and many other professionals must work together to research the topic comprehensively. International initiatives and conferences bring together experts from different areas to share their knowledge and develop new approaches. Citizen participation and community projects also play an important role in research into extraterrestrial contact. More and more people are interested in actively participating in the search for extraterrestrial life and sharing their experiences and observations.

5. Theoretical considerations and debates

In addition to the technological and scientific approaches, there are also theoretical considerations and debates. The Fermi Paradox discussion, which addresses the question of why, despite the possible existence of many potentially life-friendly planets in the universe, no clear evidence of extraterrestrial life has yet been found, is an example of this. Such debates stimulate thought and can inspire new perspectives and approaches to research.

6. Outlook into the future: contacts with extraterrestrials

Looking into the future with extraterrestrial contact offers a fascinating look at the possibilities and challenges. It is important that we continually develop our communication strategies and skills in order to be prepared for possible future contacts. Extraterrestrial contact not only opens up the opportunity to expand our understanding of the universe, but also to reflect on our own humanity and gain new perspectives on life in space.

Framework conditions

General legal aspects

In the world of interstellar relations, it is essential to explore the legal dimension. At present, it is difficult to imagine that Earth's existing laws and regulations apply to contacts with extraterrestrial life forms. To date, there are no international agreements or treaties specifically created for interaction with extraterrestrial life. If such agreements exist, their relevance is questionable. To date, there are no official reports of extraterrestrial beings that have already signed a legal framework or treaty. This fact opens up a wide field for speculation and open questions.

These questions that arise regarding the legal environment for contact with extraterrestrial life are diverse and complex. Some of these will be examined in more detail in the coming sections. Who is responsible for first contact with extraterrestrial life forms and how should this take place? How are ownership rights to resources and technologies acquired through contact with extraterrestrial life forms regulated? What regulations are necessary to establish and regulate trade relationships with alien species? What laws apply to the coexistence of humans and alien life forms in shared settlements or space stations? These questions and many more questions must follow, even if we do not yet have satisfactory answers to them.

The rights and responsibilities of the various species in the universe are of great importance. How can these rights be protected and how is the sovereignty of different species respected and maintained?

When disagreements arise between different species, the question arises as to how conflicts can be resolved. The transfer of technology and knowledge between these species must also be regulated. Last but not least, we need to consider who bears responsibility for actions that could affect the lives of other life forms in the universe. Ethics plays a primary role in this context. What ethical standards must be observed when dealing with extraterrestrial life and in intergalactic diplomacy?

You should think about these questions early on in order to ensure smooth cooperation between the different life forms in the universe and to avoid unwanted conflicts. New international agreements and laws may need to be developed that are specifically tailored to the needs and requirements of dealing with extraterrestrial life. In order to meet these challenges, humanity should prepare for these issues in good time.

The legal dimension of contact with extraterrestrial life requires careful consideration of responsibilities, rights, obligations and ethical standards. A possible solution could be the creation of an international organization or committee to deal with intergalactic law issues and serve as a platform for dialogue and cooperation.

The aspects and questions mentioned are currently still hypothetical in nature and cannot be answered conclusively based on the current state of knowledge. Nevertheless, it is of utmost importance to address these questions and prepare for potential contact with extraterrestrial life. Humanity may be facing an exciting future of intergalactic exchange and cooperation. It is our responsibility to face these challenges with foresight, understanding and respect in order to enable peaceful and harmonious coexistence with other life forms in the universe.

Rights and duties in the universe

The discussion of the rights and responsibilities of entities in the universe is inherently complex and controversial because there are no universally accepted laws or standards followed by all entities in the cosmos. Nevertheless, there are various legal and ethical frameworks that can serve as a guide for the coexistence of different forms of existence.

One approach to solving this problem is the concept of "cosmic responsibility". It postulates that all intelligent beings in the universe have a responsibility to ensure that their actions are consistent with the preservation and promotion of life throughout the universe. This responsibility extends to the protection of one's own species as well as to the well-being of other life forms and their habitats.

There are international agreements and conventions related to the protection and preservation of creatures in the universe, such as the Convention for the Protection of Outer Space and Planets. This agreement regulates the prohibition of the contamination of celestial bodies with organic and biological materials, as well as the use of resources without appropriate authorization. In addition to legal considerations, moral aspects are of great importance with regard to the protection of beings in the universe. Particularly when it comes to alien species that may not have reached the same level of development or progress as humanity, moral reflection is essential. In such situations, it is of utmost importance to exercise empathy and compassion and ensure that no harm is done. It should be noted, however, that these considerations may not apply to extraterrestrial visitors, as their level of development may be far ahead of our own due to their visit to Earth.

The question of the rights and responsibilities of creatures in the universe is a complex issue that requires legal, moral and ethical considerations. The idea of?? cosmic responsibility and the protection of beings and habitats serve as a guideline to ensure that all beings in the universe are respected and cared for.

Accountability at first contact

Who should take the lead?

Have you ever thought about who should be responsible for first contact with extraterrestrial life and how it should occur? This question raises a number of considerations, as we currently have no official authority or institution specifically responsible for dealing with extraterrestrial life. However, there are different approaches and ways you can tackle this challenge.

A possible role in responsibility for first contact could be played by governments at the global level. Because first contact is an event of enormous significance and potentially far-reaching consequences, it would make sense for governments to come together and develop a coordinated approach. International organizations such as the United Nations or the International Astronomical Union (IAU) could play a supporting role as platforms for exchanging information and coordinating actions.

However, there is also the possibility that initial contact occurs in a spontaneous and unpredictable way, such as receiving signals from space or the appearance of alien spacecraft on Earth. In such a scenario, every human would have to react flexibly and ensure that they have a clear communication strategy aimed at facilitating safe, understanding and peaceful exchanges with the extraterrestrial life forms.

Regardless of who is ultimately responsible for the initial contact, it is extremely important that it occurs in a respectful and sensitive manner. Both humanity and extraterrestrial life forms should have the opportunity to meet each other safely and without harm. Open and transparent communication based on mutual understanding, respect and cooperation should be the foundation for this historic moment. You must also take into account cultural differences and the diversity of extraterrestrial life forms and strive for intercultural exchange in order to avoid misunderstandings and strive for peaceful coexistence.

The question of responsibility during first contact with extraterrestrial life forms is undoubtedly complex. It requires careful planning, preparation and global collaboration because no one can say where or how the first encounter will take place. By adopting a responsible and ethical approach, you can ensure that this historic moment becomes a positive and enriching experience for everyone involved.

Property rights

Ownership of resources and technologies

The question of ownership rights to resources and technologies is an extremely complex and important issue that could be of central importance in the event of contact with extraterrestrial life forms. Suppose alien life forms possess valuable resources that could be of significant benefit to humanity, such as breakthrough energy sources, innovative cures, or revolutionary materials. In such a scenario, ethical and moral issues must be carefully discussed to ensure that these resources are shared and used fairly without violating the rights of the alien life forms.

It would be neither appropriate nor fair to claim these resources from alien life forms simply because of our perceived technological superiority (which may well be questioned given our past successes in space travel) or because of our larger population size. Instead, a cooperative approach should be adopted, in which humanity and extraterrestrial life forms decide together how best to use resources for mutual benefit. A fundamental question is how a fair distribution of these resources can be achieved. Principles such as justice, sustainability and the protection of the natural environment can play a crucial role. There may be a need for international agreements or treaties that establish clear rules and standards for dealing with this extraterrestrial potential.

The technologies developed by extraterrestrial life forms could also have a significant impact on humanity. These technologies could help us solve pressing problems, be it in the fields of medicine, energy production or space travel. Extreme caution is required to ensure that these technologies are used responsibly and that the rights of alien life forms are protected. It may be necessary to develop international agreements and ethical guidelines to ensure the responsible use of such technologies and ensure that they serve the well-being of all humanity. This requires deep intercultural and intergalactic collaboration to promote harmonious coexistence and address opportunities and challenges alike.

Should encounters with extraterrestrial civilizations actually occur in the future, the question of ownership of advanced technologies and resources could be of enormous interest. In such a situation, questions could arise about who has the right to use these resources and whether there should be ethical guidelines to ensure that the use of these resources is fair and sustainable.

Ownership of extraterrestrial resources and technologies is a highly complex and controversial issue that includes legal, ethical and social dimensions. There are currently no international laws or standards specifically developed for this case. It is therefore of utmost importance that the international community cooperates to develop clear policies and agreements that govern the use of these resources and ensure fair distribution. International cooperation and the exchange of expertise could play a crucial role.

With all the thoughts written in this book, there is one unknown. These unknowns are the aliens themselves. Even if governments, countries and peoples on our planet agree on how to deal with all resources and values, that does not mean that this is done in the interests of the visitors from space.

So the big question remains!

Given the advanced capabilities of extraterrestrial life forms, the key question remains how they will respond to our ideas about ownership of resources and technologies. This raises the question of whether Earth is the only planet that has been visited by aliens. If there are already other visited planets, the question arises about the rules established there – who established them and what are the consequences? The possibility that we will have to submit 100% to the regulations of extraterrestrial civilizations opens up a fascinating scenario that can also be considered in this context.

Trade relations

Possible trade relations with alien species

Establishing trade relationships with alien life forms requires careful consideration and preparation. It is crucial to take the interests and needs of both sides into account to ensure effective communication.

Clearly defined property rights are of great importance in trade transactions with extraterrestrial life forms. Rules and regulations should be established to enable fair and equitable trade that benefits all parties involved. Security also plays an important role. It is essential that traded resources and technologies are safe and do not pose a threat to the health or safety of the species involved. In addition, potential impacts on the environment and the universe as a whole should be taken into account. Trade in extraterrestrial life forms can have unforeseen consequences for ecosystems and planets, which is why it should be managed sustainably and responsibly.

The basis of all trade relations with extraterrestrial life forms should be cooperation and mutual respect. Both sides should benefit from the partnership and strive for long-term benefits. In order to regulate trade relations with extraterrestrial life forms, establishing a legal framework is a high priority. International agreements and treaties can establish clear guidelines for trade, protection of rights and conflict resolution. One possibility is to create a specialized agency responsible for regulating and monitoring trade in alien life forms.

This authority could conduct negotiations, set trading conditions and mediate disputes.

Promoting research and development in interstellar trade is an area that cannot be neglected. This includes exploring new trading opportunities, developing trading technologies and training professionals for intergalactic trade. It is crucial that trade relationships with extraterrestrial life forms are long-term and sustainable. Humanity should benefit from the experiences and resources of extraterrestrial life forms and at the same time respect the rights and needs of everyone involved. Open and transparent communication is an important aspect to avoid misunderstandings so that common goals can be defined.

It is important that both sides are allowed to express their interests and concerns. A comprehensive analysis of the economic aspects is also necessary to establish a successful trading relationship with extraterrestrial life forms. This includes assessing demand for extraterrestrial resources and technologies, identifying potential trading partners, and assessing the economic impact on the species involved. It is advisable to also introduce conflict resolution and dispute resolution mechanisms, as commercial relationships can always involve risks and disagreements. Arbitration or the involvement of a neutral mediator may be helpful here.

Furthermore, the protection of intellectual property is extremely important in trade relationships with extraterrestrial life forms. It is crucial to develop mechanisms to protect patents, copyrights and other intellectual property rights. This promotes innovation and secures the interests of all parties involved.

The trade in extraterrestrial life forms should not be viewed in isolation, but should be embedded in a broader context. Collaboration with other governments and organizations is important to develop best practices and share experiences. Establishing trade relationships with alien life forms should be based on cooperation, mutual benefit and long-term prosperity for all species involved. The issue of communication and trade relations with extraterrestrial life forms is a complex challenge that requires thorough planning, communication and collaboration. Through a comprehensive legal framework, a sustainable approach and a focus on cooperation and mutual respect, trade relationships with extraterrestrial life forms can contribute to mutual benefit and promote interstellar progress.

Living together in settlements

Since humans and alien lifeforms living together in shared settlements or stations is a very hypothetical scenario, there are currently no specific laws regulating this. However, existing laws and treaties, such as international law and international agreements, could serve as guidelines for coexistence.

Some legal questions may arise: How are ownership rights in shared resources and technologies governed? How are conflicts between different species resolved?

Who is responsible for the security and defense of the joint settlements or stations? How are criminal offenses and crimes against members of another species, but also against one's own race, punished?

To resolve these issues, it is advisable to create a common governance structure for the settlements or stations, in which representatives of both species work together and make joint decisions. This enables balanced consideration of the interests and needs of all residents. In addition, a legal framework could be developed that is specifically tailored to the needs and challenges of coexistence between humans and extraterrestrial life forms. This framework should contain clear provisions on how ownership of shared resources and technologies can be determined and how potential disputes can be resolved.

The security and defense of the shared settlements or stations would be a shared responsibility of both species. It would be important to establish mechanisms for effective cooperation and coordination to identify potential threats and respond appropriately. In the case of crimes and crimes against members of another species, a fair and transparent legal system should be created, aiming at equal treatment and justice for all residents of the settlements. This could include the establishment of specialized courts or tribunals tasked with resolving such cases. To ensure harmonious coexistence, it is essential that communication, mutual understanding and respect between different species are promoted. Educational programs and intercultural exchange could help reduce prejudices and promote peaceful coexistence.

Since coexistence in settlements of humans and extraterrestrial life forms would be challenging, extensive preparation and planning is required to address potential problems in order to enable harmonious coexistence.

Sovereignty

When it comes to extraterrestrial life, different considerations about the sovereignty of nations and states arise. Some countries might argue that they have the right to protect their borders from alien invaders since control of their territory is their responsibility. On the other hand, the argument could be that the existence of extraterrestrial life has implications for all of humanity. It is therefore necessary to take global action to ensure that everyone's interests are taken into account. This could lead to global regulation that ensures the protection of both humanity and alien species.

There are currently no international agreements or treaties that are publicly available and explicitly address sovereignty in the context of extraterrestrial life. However, this could become a subject of debate and negotiation in the future, particularly in the event of first contact or collaboration with extraterrestrial life forms.

The issue of sovereignty in the context of extraterrestrial life is extremely complex and requires careful consideration. One possible approach could be to promote international discussions and negotiations to find consensus on how to respect sovereign rights while taking into account global interests and the protection of all parties involved. One solution could be to create an international organization or council that would serve as a forum for exchange and coordination among nations. This could develop guidelines and recommendations to preserve the sovereignty of states while promoting a cooperative approach to dealing with extraterrestrial life.

In any case, it is of the utmost importance that discussions about sovereignty and extraterrestrial life be conducted on a basis of respect and understanding. The different viewpoints should be heard and weighed in order to find a fair and balanced approach that takes into account the concerns of all parties involved. This approach will be crucial in shaping the future of dealing with extraterrestrial life in a globalized world.

Conflict resolutions

Resolving conflicts between different species in the universe is undoubtedly a complex matter that requires careful consideration of many aspects. When dealing with conflicts with extraterrestrial life forms, one faces challenges that often result from the different cultural ideas and value systems of the species involved.

A possible approach to conflict resolution could be to establish intergalactic arbitration courts or arbitration boards that specifically deal with disputes between different species and work on a common legal basis. However, this approach assumes that the parties involved are willing to cooperate and understand each other.

Yet another possibility could be the development of an intergalactic treaty that establishes common principles and values? ?and serves as a guide for resolving conflicts between different species in the universe. This would serve as a framework for conflict resolution and help establish clear expectations and standards.

Effective conflict resolution requires that all species involved take responsibility and actively strive to find compromises and promote peaceful solutions. Each species can contribute its own resources and abilities.

Open dialogue and respectful communication between the parties involved are fundamental elements for successful conflict resolution.

By exchanging viewpoints, understanding each other's motives and needs, and finding solutions together, conflicts can be resolved peacefully.

When dealing with conflict, it is essential that the rights and needs of all parties involved are equally taken into account. A fair and balanced approach is of utmost importance to find sustainable solutions and promote long-term harmony in the universe. Such an approach would help conflicts to be resolved in a respectful and cooperative manner, thereby contributing to the well-being and stability of the entire cosmos.

Technology transfer

The transfer of technology between different species in the universe can be extremely complex due to possible differences in technology, culture and ethics of the species involved. It is important that the species involved agree on clear rules and procedures for technology transfer to ensure that the technology is used responsibly and that no harm is caused to the species involved or the universe.

One way to regulate technology transfer is to establish an international organization or body responsible for monitoring and regulating technology transfer between different species in the universe. This organization could develop policies and procedures for technology transfer and monitor them to ensure that the technology is used responsibly and that there are no undesirable effects on the species involved or the universe.

It is also important that the entities involved are able to use and maintain the technology safely and effectively. Education, training sessions and inter-species exchange programs can be helpful in ensuring that the technology is used and maintained responsibly. Additionally, it is important that the creatures involved consider the technology's impact on the environment and other species in the universe. A comprehensive environmental assessment can help identify potential risks and take action to minimize negative impacts on the environment and other species.

When it comes to technology transfer, open and transparent communication is extremely important. By exchanging information, sharing knowledge and learning together, potential misunderstandings or conflicts can be avoided. Technology transfer should be based on reciprocity and both species should benefit equally from collaboration. A fair distribution of benefits and a respectful recognition of the capabilities and contributions of each life form are of great relevance to ensure a long-term and sustainable partnership in technology transfer.

Responsibility

Responsibility for actions that may affect the lives of other species in the universe lies primarily with each species involved. There may be universal ethical and moral standards that apply to all species in the universe, but there is no official authority is responsible for enforcing these standards.

When a species takes an action that could affect the lives of other species, it is paramount that it is aware of the consequences and carefully considers the impact of its actions on other species. In certain cases it may be necessary for the international community or a similar organization to intervene to minimize the damage caused and prevent possible future damage.

All species in the universe should be aware that their actions can have consequences for other life forms. Each species has a responsibility to ensure that it acts in accordance with universal ethical and moral standards to protect the lives of other species in the universe.

An effective way to raise awareness of this responsibility is to facilitate the exchange of information and knowledge between species. Open and transparent communication can help avoid misunderstandings and raise awareness of the consequences of certain actions. Additionally, the establishment of an international organization or body could serve as a forum to discuss ethical issues related to extraterrestrial life and develop guidelines for responsible action. This would help promote a shared understanding of responsibilities to other species in the universe.

It should also be noted that each species takes an active role in identifying potential risks and developing strategies to minimize risks. This can be achieved by conducting environmental assessments, considering sustainable practices and adhering to precautionary measures.

As each species assumes its responsibilities and commits to the protection and well-being of other species in the universe, harmonious and respectful coexistence can be achieved. The responsibility of each individual contributes to making the future of interstellar life in the universe safe and worth living.

Ethics

Ethical standards must be adhered to when dealing with extraterrestrial life and in intergalactic diplomacy to ensure that the rights and dignity of all species involved are respected. Some of the most important ethical standards could be

Respect for the life and individual rights of all species in the universe: Each species should recognize and respect the uniqueness and value of the lives of other species. This means not only recognizing physical existence, but also taking into account cultural, religious and social differences. It also requires respect for fundamental rights and freedoms for all living beings, regardless of their origin or species.

Responsible use of resources and technologies: They must not be obtained or used at the expense of other species. This requires sustainable use of natural resources and careful consideration of the environmental impacts of technologies and developments. A species should ensure that its actions do not result in damage to the environment or ecosystem that negatively impacts other living things.

Avoidance of violence and peaceful conflict resolution: Conflicts should be resolved through dialogue, negotiation and mediation in order to promote non-violent and harmonious coexistence. This requires empathy and understanding of other species' perspectives in order to find common interests and solutions. By promoting tolerance and mutual respect, conflicts can be resolved peacefully.

Promoting collaboration and solidarity: between different species to achieve common goals. By sharing knowledge, resources and skills, species can learn from each other and work together to create a better future in the universe. This requires a willingness to work together across cultural and political boundaries and to prioritize common interests over individual differences.

Commitment to transparency and openness in communication and collaboration. Honest and transparent communication promotes trust and enables effective collaboration. This requires disclosure of information relevant to the collaboration and a willingness to speak openly about goals, concerns and expectations. Through open and trusting communication, misunderstandings can be avoided and common goals can be achieved more effectively.

These ethical norms can help ensure that the treatment of extraterrestrial life and intergalactic diplomacy are conducted in a morally responsible and sustainable manner. If everyone respects and acts according to ethical principles, a fair and harmonious coexistence between species in the universe can be achieved.

Spiritual and philosophical reflections

Encountering extraterrestrial life can raise profound spiritual and philosophical questions that challenge our understanding of the human place in the universe and our relationship to other intelligent beings. Some people see the discovery of extraterrestrial life as an opportunity to expand and rethink our spiritual beliefs and perhaps even change our understanding of God and religion. They see this as an invitation to deepen our faith and place it in a larger cosmic context.

On the other hand, some argue that the discovery of life outside Earth forces us to think about our own humanity in a larger universe. This encounter could help us see our existence in a new light and understand our connection to other intelligent beings on a larger scale.

There are discussions in philosophy about how the discovery of extraterrestrial life might influence our understanding of ethics and morality. Encountering intelligent extraterrestrial life could raise questions about how we treat these beings and what responsibility we have for their lives and rights. This could lead to a reassessment of our ethical principles and cause us to reconsider our moral obligations to other intelligent beings.

All of these spiritual and philosophical considerations underscore the need to prepare for the possibility of discovering extraterrestrial life. We must be prepared to adapt to new and challenging questions and scenarios. It is important to remain open and see this encounter as an opportunity to expand our knowledge and gain new insights about ourselves and the universe. The discovery of extraterrestrial life could broaden our spiritual and philosophical horizons and inspire us to further explore the profound questions of our existence in the cosmos.

Living with aliens

Living together

Visions and reality

Possible coexistence with extraterrestrial life forms is a theme explored in many science fiction stories and films. In reality, however, it is difficult to predict what such coexistence would actually look like. It depends on a variety of factors, such as the physiological, biological, cultural and technological differences between species.

However, some authors and scientists have outlined possible scenarios of what coexistence with extraterrestrial life forms could look like. An example of this is a type of intergalactic diplomacy in which representatives from different planets meet regularly and discuss common concerns, such as conserving resources or resolving conflicts. Another possibility would be for humans and extraterrestrial life forms to live together in special space stations or colonies and work together to improve their living conditions and achieve common goals. In such a scenario, it would be important that both humans and alien life forms are willing to compromise and respect each other's cultural differences and needs. However, there are also potential challenges and risks to coexisting with extraterrestrial life forms. For example, conflicts could arise when resources are scarce or when the interests of both species conflict. It might also be difficult to find a common language or bridge cultural differences.

Another risk is that diseases and viruses could spread between species because each other's immune system is unknown. Therefore, it is important to take strict safety precautions to minimize the risk of illness and infection when living with extraterrestrial life forms.

It is difficult to predict what coexistence with extraterrestrial life forms would actually look like. Nevertheless, it is important that humanity prepares for a possible encounter with extraterrestrial life forms. This includes improving knowledge of physiology, biology, diplomacy and technology, as well as preparing for possible challenges and risks that could arise from such coexistence.

Intergalactic coexistence

The challenges of intergalactic diplomacy

Intergalactic diplomacy is a captivating field that deals with diplomatic relations between different civilizations in the universe. The focus is on the exchange of information, technologies and resources, the settlement of conflicts and the creation of peaceful relationships between civilizations.

This discipline presents many challenges that must be overcome. In particular, these different civilizations should be able to communicate with each other even if they speak different languages? ?and have different cultural backgrounds. In addition, it is essential that they approach each other and have the ability to find compromises, even if their interests and needs differ.

Resolving conflicts is another important aspect of intergalactic diplomacy. Should conflicts arise between civilizations, it is of utmost importance that they be resolved peacefully to prevent escalations and possible acts of violence. Mediators play an important role here as they mediate between the affected parties and try to find a solution that is acceptable to all parties. Intergalactic diplomacy is closely linked to intergalactic legislation. Within this framework, common rules and laws are developed that apply to all civilizations in the universe and are aimed at peaceful coexistence and cooperation.

Overall, intergalactic diplomacy is a complex and demanding discipline that presents both challenges and opportunities. If the different civilizations in the universe are able to reach out to each other, communicate and work together, they can learn from each other and work together to create a better future for everyone.

It is of the utmost importance that the people involved in this important task are not driven by greed for profit, but are primarily committed to communication, coexistence and the peaceful coexistence of different forms of life. Because anything else would not work and could lead to conflicts, arguments and possibly even wars. Given our current inability to visit living beings on another planet, we are undoubtedly the ones who would lose such a confrontation.

The crucial question is: Do we want that? Peaceful coexistence remains the best basis for a harmonious future.

Space colonies

The establishment of special space stations or colonies offers a promising possibility for humans and extraterrestrial life forms to coexist. This could be a joint effort to create the conditions for survival and cooperation in a shared environment. Such facilities would allow different species to work together and pursue common goals.

However, to create a successful collaboration, both humans and alien life forms must be willing to compromise and respect each other's cultural differences and needs. It is necessary to understand the social and cultural conditions that enable successful collaboration. This includes exchanging information, creating trust and developing common goals.

Such collaboration would also generate new technologies and scientific knowledge, as diverse species would benefit from their experiences and knowledge. It would also provide the opportunity to exploit the universe's resources and open up new habitats.

However, there are also risks with such collaboration. Conflicts and misunderstandings could arise when cultural differences and prejudices are not understood or respected. Communication problems could arise if the language or other methods of communication are not understood.

Successful collaboration between humans and extraterrestrial life forms therefore requires careful management of the challenges and opportunities that arise from this unique situation. It requires the ability to be open and to embrace new experiences and perspectives in order to enable lasting and successful collaboration.

The creation of special space stations or colonies where humans and alien life forms work together to improve their living conditions is another possibility for coexistence. Such stations or colonies could be established in suitable areas of space and provide necessary resources.

In this scenario, humans and alien life forms must be willing to compromise to respect their cultural differences and needs. Creating a common culture and language could be an important step to ensure smooth collaboration. Rules and guidelines for coexistence should also be established to avoid conflict and ensure the safety of both species.

Collaboration on space stations or colonies offers a variety of benefits, including the exchange of knowledge and technology, the joint discovery of new knowledge, and the possible reduction of inter-species conflict on Earth. However, there are also challenges to be overcome, such as cultural differences, language barriers and adapting to the living conditions in such facilities.

Ensuring the safety of both species is of paramount importance, which includes incorporating specific measures to prevent threats or harm. Policies should also be established to deal with possible threats from space or other extraterrestrial species.

Successful collaboration between humans and extraterrestrial life forms requires a careful approach to the challenges and opportunities that arise from this unique situation. A willingness to accept and respect differences is essential, while at the same time creating a common basis for communication, cooperation and conflict resolution.

Additionally, such collaboration could have a positive impact on the development of respective societies, as the exchange of knowledge, technology and ideas could lead to new perspectives and an expanded understanding of themselves and the universe. However, it is crucial to consider ethical issues that respect the rights and needs of both species and create mechanisms to prevent abuse.

Overall, the idea of?? collaboration between humans and extraterrestrial life forms in space stations or colonies offers numerous opportunities and challenges. It is of utmost importance that both humans and extraterrestrial life forms are willing to compromise and treat each other with respect to ensure successful and sustainable collaboration. Only through open communication, understanding and collaboration can we shape a shared future in the universe.

Challenges and opportunities

Coexistence with extraterrestrial life forms is a fascinating prospect, but one that also presents some challenges and risks. One way to minimize these risks is to establish common laws and rules. This could ensure that both species are treated fairly and their needs and interests are adequately taken into account. This would create a basis for harmonious coexistence and reduce potential conflicts.

Another important aspect that must be taken into account when coexisting with extraterrestrial life forms is the question of biological compatibility. It is possible that the biological differences between species are so significant that they cannot survive in the same environment. In such a case, special adjustments and precautions would have to be taken to enable successful coexistence.

What is evident is that humanity may not be the most technologically advanced species in the universe. There could be extraterrestrial life forms that are far superior in terms of technology, knowledge and skills. In such a scenario, we would have to adapt flexibly to the new circumstances and prepare for the challenges that arise from such a constellation.

It can be seen that living together with extraterrestrial life forms involves both opportunities and risks.

Humanity should prepare for the possible impacts of contact with extraterrestrial life by developing strategies and protocols to successfully deal with these challenges. At the same time, we should focus our attention on the opportunities that could arise from collaboration with other species in the universe. This opens up the opportunity to expand our knowledge to shape a shared future in space.

Diseases and viruses

A Challenge in Alien Contact

When it comes to contact with extraterrestrial life, we must not only focus on the exciting prospects and opportunities, but also consider the potential risks and challenges. One of the critical questions that arise when living together and collaborating with extraterrestrials is the possible transmission of disease.

The biological differences between our species and extraterrestrial life forms could be significant. This includes not only differences in anatomy, but also in the immune system and biological processes. These differences could mean that pathogens that are harmless to one species could be devastating to another species. Throughout human history, we have learned how susceptible we are to infections, be they bacteria, viruses or other pathogens. The idea that we could be exposed to potentially dangerous pathogens if we come into contact with extraterrestrials is disturbing.

The challenge is not only to identify the risks, but also to develop measures to minimize these risks.

Here are some possible approaches:

Quarantine and Isolation: Before direct contact between humans and extraterrestrials occurs, it may be necessary to develop and apply quarantine measures and isolation protocols. This would help prevent the possible spread of diseases and viruses while also allowing time to study biological differences.

Vaccine development: Research and development of vaccines that protect both humans and aliens could be a game-changer. This requires a deep understanding of the biological differences and similarities between species.

However, with regard to the health sector, the following question arises again: if there are beings capable of visiting other planets, their research is undoubtedly much more advanced than ours. It is unlikely to assume that our current efforts to develop vaccines, whether for aliens or ourselves, will have a significant effect at this point. If these beings are able to traverse galaxies, it is certain that they have advanced methods to protect their own health.

It will undoubtedly take decades before humanity is able to develop vaccines against potential diseases from space. In the meantime, we can only hope that extraterrestrial beings visiting us already have advanced health protection measures and technologies in place to minimize the risk of disease transmission.

Extraterrestrial visits

Speculation and discussions

Myth or reality?

There has been speculation and discussion about possible extraterrestrial visitors to Earth for decades. Many people claim to have witnessed UFO sightings and extraterrestrial encounters, while others dismiss these stories as pure fantasy or misunderstandings. Unfortunately, scientific evidence or clear facts regarding extraterrestrial visitors are not publicly available.

Proponents of an extraterrestrial presence argue that there are numerous testimonies from credible individuals reporting encounters with extraterrestrial life forms. They point to unexplained phenomena such as UFO sightings, abduction reports and mysterious circular formations in fields. These people believe that extraterrestrial civilizations have already visited Earth and may even have made contact with the human population.

In contrast, skeptics argue that most UFO sightings and encounter reports are due to natural phenomena, optical illusions or misinterpretations. They emphasize that there is no verifiable evidence for the existence of extraterrestrial life forms and that eyewitness testimony alone is not enough to support these claims. The scientific community does not recognize conclusive evidence of extraterrestrial visitation or contact on Earth.

Still, searching for extraterrestrial life is an active area of?? research that includes, among other things, the study of exoplanets and the analysis of radio signals from space. In the face of this controversy, it is advisable to maintain an open and critical attitude. It is important to examine information carefully and not jump to conclusions. Everyone should form their own beliefs based on available evidence and scientific knowledge.

The search for intelligent extraterrestrial life is a focus of astronomy and astrobiology. Researchers are actively searching for signs of extraterrestrial life in the universe. The discovery of exoplanets located in the habitable zone around their stars has increased hope for the existence of life-friendly environments. Some argue that the diversity of the universe makes it likely that there are other planets capable of living. However, others point out that the emergence of life is a complex process that may occur rarely.

The question of contacts and communication with extraterrestrial civilizations is also controversial. There are claims that SETI (Search for Extraterrestrial Intelligence), the search for extraterrestrial life, has already received potential signals or messages from space. However, such interpretations are often questioned. The possibility of misunderstandings or different forms of communication between different life forms could also be discussed.

Another controversy arises due to discussion of possible cover-ups and secrecy by governments. Some believe that governments around the world are withholding information about extraterrestrial visits to avoid panic or other negative effects on society. Others argue that such conspiracy theories are baseless and that governments have no evidence of extraterrestrial contact.

These controversies have been and will continue to be the subject of discussion and research. Since there is no conclusive evidence of extraterrestrial contact, it is advisable to maintain a critical and rational approach.

Nevertheless, the question arises as to why governments keep images and witness statements under wraps for 50 years if there is supposedly no extraterrestrial contact. Equally puzzling is the fact that high-ranking military officers are threatened with prison if they break this secrecy. This lack of transparency is incomprehensible to many people. One thing is certain, however: the more secrets are made, the more intensively they are discussed and researched. Openness to new information and a respectful examination of different viewpoints enable a well-founded discussion about this fascinating topic.

Eyewitnesses, UFO sightings

The Project Blue Book (1952-1969)

"Project Blue Book" was an actual US Air Force investigative project that dealt with UFO sightings and reports of unidentified flying objects. The project was launched in 1952 and lasted until 1969. During this time, "Project Blue Book" investigated and documented several thousand UFO sightings and incidents. I would like to present this project in a case study as it is one of the most well-known and extensive programs investigating UFO sightings and extraterrestrial encounters. Additionally, I intend to use it as an example to illuminate how governments approach the collection, investigation and assessment of such reports.

Project Blue Book

Background:
"Project Blue Book" was an American investigative project launched by the US Air Force in 1952. Its main goal was to collect, investigate and classify UFO sightings and reports of unidentified flying objects. During its existence, thousands of such reports were submitted by citizens, pilots, military personnel and other sources.

Goals and approach:

The goals of "Project Blue Book" were diverse. It should be determined whether UFOs pose a potential threat to national security and an attempt should also be made to clarify the nature and origin of the reported UFOs. The project employed astronomers, engineers, psychologists and other experts to analyze the reported sightings.

Results and conclusions:
During its existence, "Project Blue Book" investigated thousands of UFO reports. Most of these reports have been identified as natural phenomena, optical illusions, balloons or misinterpretations. Only a small number of cases remained classified as "unidentified", meaning no conclusive explanation could be found. However, the project concluded that UFOs did not pose a threat to national security and that there was no convincing evidence of extraterrestrial visitors.

Criticism and controversy:
Project Blue Book has often been accused by critics of covering up reports and not being objective. Many UFO enthusiasts believed that the US government was withholding information about extraterrestrial contacts. The closure of "Project Blue Book" in 1969 added to the ongoing controversy.

Meaning:
"Project Blue Book" remains an important case study in the history of UFO investigation. It shows how govern-

ments have handled reports of UFO sightings and extraterrestrial encounters. The discussion surrounding the disclosure of secret government documents and the investigation of UFO incidents continued for decades and remains a topic of great interest to the public today.

The Majestic Twelve (formed 1947)

Background:
The Majestic Twelve, or MJ-12 for short, is a secret group of scientists, military personnel, and government officials said to have been formed in 1947 after the Roswell Incident. President Harry S. Truman is said to have created it through an executive order. Its main purpose was to study and manage the phenomenon of extraterrestrial presence on Earth. The Roswell Incident refers to the suspected crash of an alien spacecraft in Roswell, New Mexico. The existence of MJ-12 was first claimed in 1984 through leaked documents, which has led to ongoing controversy and speculation about its authenticity.

Goals and approach:
 MJ-12 was intended to initiate and ensure communication and cooperation with extraterrestrial civilizations. The group was tasked with collecting, classifying, and protecting all information about UFO sightings, alien spacecraft crashes, and alien technology. The members of MJ-12 were sworn to strict secrecy and had access to highly classified information. It is speculated that MJ-12 was also

responsible for studying extraterrestrial biology and possible threats from extraterrestrial civilizations. The exact functioning and especially the composition of MJ-12 have long been the subject of controversy and speculation.

Results and conclusions:
The authenticity of MJ-12 and its associated documents remains the subject of intense controversy to this day. Proponents argue that the leaked papers are genuine government documents and confirm the group's existence. Skeptics, however, claim that the papers were forged and are part of a broader disinformation campaign. The US government has always denied that MJ-12 exists. However, there is no conclusive evidence that MJ-12 ever existed or still exists. Most researchers and scientists view the claims about this secret group with skepticism, citing the lack of hard evidence. It is important to acknowledge the discrepancies and different viewpoints regarding MJ-12 and to continue to seek new information and official statements.

Criticism and controversy:
The controversy surrounding MJ-12 largely revolves around the authenticity of the leaked documents and allegations that members of the group maintained strict secrecy. The US government has denied the existence of MJ-12 and claimed that the documents are forged. The persistence of politics over decades in denying something that supposedly does not exist is a fascinating and often

frustrating matter. Especially in the early third millennium, when technological advances enable unprecedented levels of surveillance and documentation, it is becoming increasingly difficult to deny things that have already been recorded.

Meaning:

The story of MJ-12 remains a fascinating chapter in the world of UFO and conspiracy theories. It illustrates the ongoing fascination and suspicion surrounding the possible cover-up by governments of information about extraterrestrial contacts. MJ-12 serves as a case study in how conspiracy theories can arise and persist despite a lack of evidence. But there is also no denying that The Majestic Twelve remains a controversial and mysterious chapter in the world of UFO and conspiracy theories. Despite ongoing controversy and speculation, there is no convincing evidence of the existence of this secret group. It remains a fascinating example of how secrecy and conspiracy theories can work in the world of UFOs and extraterrestrial encounters. Finally, it should be noted that the discussion about MJ-12 continues to be intense. The lack of clear evidence and the controversy surrounding the leaked documents leave room for speculation and different opinions. It is up to each individual to evaluate the available information and draw their own conclusions. Given the constant advances in technology and research, it is to be hoped that future discoveries and insights can contribute to further clarification. It is important to be open to new information and continue to think critically to gain a more complete understanding of the possible extraterrestrial presence on Earth. It is time for each individual to think about it and draw their own conclusions.

Who are the MJ-12?

It is extremely strange that this organization is persistently denied. At the same time, all 12 people are more or less associated with MJ-12. If you are interested in this topic, you have to expect that you will receive either no information or only very sparse information. However, one should also be comfortable knowing that there is a "no known connection to MJ-12" statement.

The question that arises here is: If there is no connection, why is there so much emphasis on emphasizing that there is no connection?

The persistence of politics over decades in denying something that supposedly does not exist is a fascinating and often frustrating matter. Especially in the early 3rd millennium, when technological advances enable unprecedented levels of surveillance and documentation, it is becoming increasingly difficult to deny things that have already been recorded.

Photos, videos and much other evidence of phenomena such as UFOs/UAPs are now ubiquitous. This evidence cannot simply be ignored or dismissed as fabrication. The wealth of public information and the ever-increasing amount of new knowledge make it increasingly unlikely that important aspects of these issues can continue to be kept under wraps.

It seems that politicians and certain government circles often have an interest in denying or downplaying the existence of such phenomena. But the more evidence comes to light and the more people become interested in these issues, the harder it becomes to maintain this stance.

There is even speculation about secret organizations such as MJ12 that are said to exist or have existed and that may have a direct impact on policy and management of UFO/UAP information. The increasing amount of information and evidence could ultimately lead to such secrets coming to light and the official position on these issues having to change.

The Majestic Twelve:

Lloyd Berkner – Ph.D. in physics
 Birthday: August 1, 1905 – June 4, 1967
 Education: BS in Electrical Engineering and Ph.D. in physics
 Work: Advisor on scientific and technical issues, including for the US government.

MJ-12 connection: There is no officially known connection between Lloyd Berkner and MJ-12.

Detlev Bronk – MD and Ph.D. in physiology
Birthday: August 13, 1897 – November 17, 1975
Education: MD and Ph.D. in physiology
Work: President of Rockefeller University, science advisor for the US government.
MJ-12 connection: There is no officially known connection between Detlev Bronk and MJ-12.

Vannevar Bush – Ph.D. in engineering
Birthday: March 11, 1890 – June 28, 1974
Education: BS in Electrical Engineering and Ph.D. in engineering
Work: Scientist and engineer, director of the Office of Scientific Research and Development during World War II.
MJ-12 connection: There is speculation that Vannevar Bush may have been connected to MJ-12, but no direct evidence of this.

James Forrestal – JD in Law
Birthday: February 15, 1892 – May 22, 1949
Education: BA in Economics and JD in Law
Job: US Secretary of the Navy, first US Secretary of Defense.

MJ-12 connection: There are claims that Forrestal was aware of UFOs and possibly MJ-12. However, there is no clear evidence of this.

Gordon Gray – BA in Law
Birthday: May 30, 1909 – November 26, 1982
Education: BA in Law
Work: US Army officer, Secretary of the Army, Director of the Office of Defense Mobilization.
MJ-12 Connection: There is no officially known connection between Gordon Gray and MJ-12.

Roscoe H. Hillenkoetter – BS in Mechanical Engineering
Birthday: May 8, 1897 – June 18, 1982
Education: BS in Mechanical Engineering
Job: First Director of the CIA.
MJ-12 connection: Hillenkoetter is often associated with UFO and MJ-12 conspiracy theories, but there is no concrete evidence of his involvement.

Jerome Clarke Hunsaker – BS in Electrical and Aeronautical Engineering
Birthday: August 26, 1886 – September 10, 1984
Education: BS in Electrical Engineering and Aeronautical Engineering
Work: Aeronautical engineer, professor at MIT.
MJ-12 Connection: There is no officially known connection between Jerome Clarke Hunsaker and MJ-12.

Donald H. Menzel – Ph.D. in astrophysics
Birthday: April 11, 1901 – March 14, 1976
Education: Ph.D. in astrophysics
Work: Professor of astrophysics at Harvard University.
MJ-12 connection: Menzel was a known skeptic of UFO phenomena and reportedly had no connection to MJ-12.

Robert M Montague:
Birthday: August 7, 1899 – February 20, 1958
Education: United States Military Academy
Job: Three-star general, commander of Sandia Base
MJ-12 Connection: There is no officially known connection of Robert M. Montague to MJ-12.

Sidney Souers – BA
Birthday: March 30, 1892 – January 14, 1973
Education: BA in Law
Job: First Director of the Central Intelligence Group (predecessor of the CIA).
MJ-12 connection: There is speculation that Souers may have been involved in UFO research, but no clear evidence of a connection to MJ-12.

Nathan F. Twining – BS
Birthday: October 11, 1897 – March 29, 1982
Education: BS in Mechanical Engineering
Job: US Air Force officer, chief of the Air Force Material Command.

MJ-12 connection: Twining is often mentioned in UFO circles, but there is no clear evidence of its involvement with MJ-12.

Hoyt Vandenberg – BA
Birthday: January 24, 1899 – April 2, 1954
Education: BA in Electrical Engineering
Job: US Air Force officer, Second Chief of Staff of the Air Force.
MJ-12 Connection: There is no officially known connection between Hoyt Vandenberg and MJ-12.

Kecksburg UFO incident (1965)

Background:
The Kecksburg UFO incident occurred on December 9, 1965 in the small town of Kecksburg, Pennsylvania, USA. It was an incident in which an unidentified flying object fell from the sky and landed in a forest area. Eyewitnesses reported that the object had a metallic, acorn-shaped structure and was decorated with hieroglyphs or symbols.

Course of the incident:
The incident began when several Kecksburg residents saw a burning fireball in the sky that appeared to have crashed around 4:45 p.m. Shortly thereafter, there was an explosion that was heard in surrounding communities. Local authorities were alerted and a search began to determine the origin of the crash. A group of eyewitnesses,

including journalists and rescue workers, reached the scene of the crash in a forest near Kecksburg. They found a metal, acorn-shaped object stuck in the ground and emitting smoke. The hieroglyphs on the object were described by some witnesses.

Authorities reaction:
Both the police and the military quickly responded to the scene. The residents of Kecksburg and the eyewitnesses were pushed back by the authorities, and the object was removed by soldiers. The explanations given by authorities varied over time. Initially it was reported that it was a "flame crashing aircraft", but this explanation was later revised. The military stated that it was actually an "unidentified flying object" but did not pose a threat.

Criticism and controversy:
The Kecksburg incident has sparked controversy and speculation for decades. Many UFO enthusiasts and conspiracy theorists claim that the object was an alien spacecraft and that the US government withheld information. There were also reports of unusual activity by men in black suits asking questions of witnesses after the incident. The expression "Man in Black" has since become a household word.

Meaning:
The Kecksburg UFO incident remains a well-known example of an incident involving UFOs and extraterrestri-

al encounters. It shows how such incidents are handled by the authorities and perceived by the public. The ongoing controversy surrounding this incident also highlights how conspiracy theories and speculation can emerge and persist in the world of UFO research.

Result:

The Kecksburg UFO incident remains a mysterious incident in which an unidentified flying object crashed in a small village and raised many questions. Despite the speculation and controversy, to date there is no conclusive evidence as to the true nature of this event. The case study makes it possible to examine the incident in the context of the discussion on extraterrestrial encounters and to present the perspectives, conclusions and controversies related to this incident.

Rudloe Manow (1974)

Background:

Rudloe Manow is a small hamlet near Corsham, Wiltshire, England. It is located near the secretive and heavily secured Ministry of Defense of Great Britain, also known as the Ministry of Defense Main Building. Apart from cameras, the entire area is secured by a massive barbed wire fence and additional military police.

It is inexplicable why this area is still so strictly guarded today when investigations are supposedly no longer taking place.

This is about a building in which there are underground passages and facilities, similar to Area 51. And Pre Astronautika suspects that extraterrestrial technologies and possibly even extraterrestrial creatures are being studied in this area. Over the past few decades, Rudloe Manow has been linked to various UFO and conspiracy theories, particularly those related to alleged Ministry of Defense activities.

Reports of UFO sightings:
There are reports from residents and visitors to the area who claim to have seen UFOs over Rudloe Manow and the surrounding area. Some of these reports suggest that the UFO sightings may have occurred near the Department of Defense. These sightings are largely based on eyewitness accounts and are not supported by reliable scientific evidence.

Conspiracy theories:
Rudloe Manow is often depicted in conspiracy theories as a location where the Ministry of Defense supposedly stores UFO wreckage and other evidence of extraterrestrial activity. Some theories claim that the British military conducted secret research into alien technology. However, there is no public evidence to support these claims.

Ministry of Defense and Rudloe Manor:
The Ministry of Defense has used the hamlet of Rudloe Manow and associated Rudloe Manor in the past. Howe-

ver, the ministry has repeatedly stressed that it has no extraterrestrial contacts or UFO wreckage in its possession. Officially, Rudloe Manor was used primarily for document storage and emergency response coordination.

Conclusion:

Rudloe Manow is a place often associated with UFO and conspiracy theories. Despite the numerous reports and theories, there is no convincing evidence of extraterrestrial activity or UFO sightings in this area. However, the connection to the Ministry of Defense and the secrecy surrounding this facility have further fueled speculation. You should approach places like Rudloe Manow with a healthy mix of curiosity, skepticism and interest. Maintain a critical approach and rely on reliable sources and evidence. The story and controversy surrounding Rudloe Manow is an example of how rumors and conspiracy theories regarding extraterrestrial encounters can arise even in the absence of concrete evidence.

Malmstrong nuclear missile crisis (1967)

The Malmstrom nuclear missile crisis of March 1967 was a notable event in which several nuclear-tipped intercontinental ballistic missiles at Malmstrom Air Force Base in Montana, USA were disrupted. In connection with this incident, there were claims that UFOs were capable of deactivating all nuclear weapons. This case study examines the events of the Malmstrom nuclear missile crisis and the claims of potential UFO intervention.

Background:
Malmstrom Air Force Base was a strategically important installation during the Cold War that housed intercontinental ballistic missiles with nuclear warheads. On March 16, 1967, unusual incidents began with several missiles jamming on the base. These incidents sparked speculation and suspicion about possible extraterrestrial influence.

Events of the Malmstrom Nuclear Missile Crisis:
On the evening of March 16, 1967, alarms were sounded at Malmstrom Air Force Base. Security personnel and maintenance teams were dispatched to determine the cause of the disruption.

During these investigations, several reports of sightings of unidentified flying objects (UFOs) were reported near the base.

Particularly notable was a report of a large red object hovering over the base's entrance gate. It was claimed that after this object was sighted, one control panel after another in the missile launch facilities were shut down. These events attracted widespread attention and led to speculation about the potential intervention of UFOs and their ability to disable nuclear weapons.

Reactions and confidentiality:

High-ranking officials at Malmstrom Air Force Base who witnessed these incidents were allegedly made to sign a non-disclosure agreement. It was claimed that this clause prevented them from speaking about the events and sharing their observations with the public. However, the exact details and extent of this secrecy are not fully known.

Discussion about UFO interference and deactivation of nuclear weapons:
The claim that UFOs are capable of deactivating all nuclear weapons is based on the events of the Malmstrom nuclear missile crisis. However, there has been no official confirmation of this claim. Investigations and analyzes of the incident have not yet provided a clear explanation for the missile malfunctions.

There are various theories and hypotheses to explain these incidents. Some suspect technical problems that may have led to the observed phenomena. These include, for example, disruptions in the communication systems of nuclear missiles or malfunctions in the control panels. Another theory is that electromagnetic interference from external sources, such as unidentified flying objects, may have caused the shutdowns.

There was also speculation about extraterrestrial influence. Some believe that the UFOs observed were capable of deliberately deactivating the nuclear weapons, either through advanced technologies or by manipulating energy fields. This hypothesis is based on testimony from high-ranking employees who claim to have seen a large red object above the front gate before the shutdowns occurred. It was believed that these testimonies were not made public due to the confidentiality clause. In any case, it is noteworthy that there is no clear evidence of extraterrestrial influence or other technical problems that could explain the Malmstrom nuclear missile crisis. The exact causes remain controversial and the subject of speculation to this day. The incident remains an interesting and controversial case related to UFO sightings and potential impacts on military installations.

The Malmstrom nuclear missile crisis has led to increased attention on the phenomenon of UFO sightings and their potential impact on security-related facilities. It has also sparked a broader discussion about the existence of extraterrestrial life and its influence on human civilization. Researchers, UFO enthusiasts and the public continue to study this case to gain more clarity about the events at Malmstrom and to find possible explanations.

Roswell Incident (1947)

The Roswell Incident in July 1947 near the city of Roswell, New Mexico is an unexpected incident in the history of UFO sightings. This event sparked a wave of speculation, conspiracy theories and investigations. This case study examines the Roswell incident and its complex aspects.

Events:

On July 2, 1947, an unidentified flying object crashed on farmer Mac Brazel's land. Local authorities became aware of the incident and the Air Force was called in. A press release that spoke of the recovery of a "flying disk" was initially published. However, this explanation was later revised and the crashed object was identified as a weather balloon.

Controversy and conspiracy theories:

After the Roswell incident was officially declared a weather balloon, rumors and speculation arose about a possible government cover-up. Conspiracy theorists claimed that the crashed object was actually an alien spacecraft and that the government was trying to keep the existence of extraterrestrial life a secret. These theories led to intense discussion and further investigation into the incident.

Witness statements and eyewitness reports:

Over the years, various witnesses and eyewitnesses have come forward claiming to have seen alien wreckage, strange creatures, and secret military activity related to the Roswell Incident. These statements led to further speculation and investigation to reveal the truth behind the events at Roswell.

In connection with the Roswell Incident, there are also reports of witnesses claiming to have seen unusual creatures or corpses.

A nurse named "Naomi Self" is sometimes mentioned in such reports. She reportedly claimed to have seen a creature with three fingers. However, there is little verifiable information about Naomi Self, and some aspects of her story are highly controversial.

Some sources claim that "Naomi Self" was never seen again after her alleged statement. This claim adds to the myth and secrecy surrounding the Roswell Incident. However, there is no verifiable evidence for the disappearance of "Naomi Self" or for her alleged statements.

Official investigations:

In the years following the incident, the US Air Force conducted various investigations to clarify the incident. The most famous was the 1994 "Roswell Report," in which the Air Force re-investigated the incident and concluded that the crashed object was actually a spy balloon as part of Project Mogul. However, this explanation could not completely dispel all the doubts and questions of critics and conspiracy theorists.

Impact and significance:

The Roswell Incident has had a significant impact on UFO research, popular culture, and public perception of extraterrestrial life. It has led to increased attention to UFO sightings and potential government cover-ups. The incident has also reignited debate about the existence of extraterrestrial life and the possibility of extraterrestrial visits to Earth. For some, the Roswell Incident is important evidence of the existence of extraterrestrial intelligence and government cover-up. On the other side are skeptics and critics who seek alternative explanations for the incident. They argue that much of the hype surrounding Roswell is based on misunderstandings, misinterpretations and sensationalist reporting. They emphasize the importance of facts and scientific explanations, such as Project Mogul's weather balloon, to explain the incident. To them, the Roswell incident is merely an example of mass hysteria and conspiracy.

Summary:

The Roswell Incident of 1947 remains a fascinating and controversial event to this day. The various theories, testimonies and investigations have led to an ongoing debate about the existence of extraterrestrial life and the cover-up by government agencies. Although official statements dismissed the incident as a weather balloon, many people maintain the belief that Roswell is evidence of extraterrestrial life. The incident has fueled debate about UFOs and extraterrestrial visits to Earth and continues to influence popular culture and public perceptions of extraterrestrial life. Extraterrestrial technology is being researched at Area 51. A lot of people are of this opinion. Because as soon as you get close, armed men in camouflage suits threaten you with weapons. What is really being guarded? A weather balloon?

Phoenix Lights (1997)

The Phoenix Lights are a well-known UFO phenomenon that occurred over the city of Phoenix, Arizona in March 1997. The incident attracted worldwide attention and led to an ongoing debate about the nature of the lights observed and their origin. This case study takes a closer look at the events of the Phoenix Lights and discusses various theories and explanations.

Background:
On the evening of March 13, 1997, thousands of people in the Phoenix area observed unusual light phenomena in the sky. The lights were in a V formation and moved slowly across the sky. Witnesses described them as huge, triangular objects or flying saucers. The sightings lasted several hours and were photographed and videotaped by many people.

Witness statements and evidence:
There are numerous testimonies from citizens, including police officers and pilots, who observed the lights. Many of the witnesses reported that the objects moved silently and were impressive in size. There is also a variety of photo and video material that documents the light phenomena.

Official statements:
Authorities in Phoenix initially reacted cautiously to the reports, explaining the lights as flares or signal munitions

fired during military exercises. However, these explanations could not explain all the observations and the impressive size of the lights. Authorities later changed their statements and explained that they were Bengal flares dropped from aircraft over the Barry M. Goldwater Range training area.

Alternative explanations and UFO theories:
Despite the official explanations, there are numerous alternative theories and speculations about the Phoenix Lights. Some UFO enthusiasts and researchers believe the lights are alien spacecraft or advanced military technology. They argue that authorities deliberately covered up the incident to hide the existence of extraterrestrial intelligence. Critics, however, see the Phoenix Lights as a product of mass hysteria, misinterpretations or atmospheric phenomena.

Effects and Aftermath:
The Phoenix Lights sparked widespread media coverage and led to increased attention around UFO phenomena. The incident has also fueled debate over the disclosure of government documents on UFOs and extraterrestrial life. Memorial events are held annually in Phoenix where people come together to talk about their observations and experiences.

Conclusion:

The Phoenix Lights continue to hold a fascination for UFO enthusiasts and researchers around the world. Despite the official explanations, many questions and ambiguities remain. The different testimonies and the extensive visual material cast doubt on the official explanations and leave room for alternative interpretations. The Phoenix Lights incident also has far-reaching implications for awareness of UFO phenomena and the disclosure of government information. It has contributed to the topic being discussed more seriously and to calls for greater transparency. The Phoenix Lights therefore represent an important case that continues to be the subject of research and discussion. They serve as an example of the complex and fascinating phenomenon of UFO sightings and encourage further search for answers to the question of extraterrestrial life and technological presence.

It remains to be seen whether future investigations and developments will provide new insights into the Phoenix Lights incident and whether these can ultimately lead to a final clarification of what happened. Until then, the case will continue to pique the interest and curiosity of people worldwide and advance the debate about extraterrestrial life and UFOs.

Vorfall in Rendlesham Forest (1980)

The Rendlesham Forest UFO incident is one of the most famous UFO incidents in history. It occurred in 1980 near RAF Woodbridge and RAF Bentwaters, two military bases near Ipswich, Suffolk, UK. During this incident, Colonel Charles Halt, the Deputy Commanding Officer of RAF Bentwaters, played a central role. This case study examines the Rendlesham Forest UFO incident and the explanations given by Colonel Charles Halt.

Events:

On December 26, 1980, several members of the military and security personnel near RAF Woodbridge observed unusual lights in the sky over Rendlesham Forest. These lights were described as pulsating, bright and colorful. The Deputy Commanding Officer of RAF Bentwaters, Colonel Charles Halt, and his security personnel also observed the lights. In the following days, on December 27 and 28, further sightings of unusual lights were reported in Rendlesham Forest. Colonel Halt and his security staff went into the forest to identify the source of the lights. They reported a triangular, shiny metallic object lying on the ground and leaving trails. Colonel Halt recorded an audio message in which he documented the events.

Statements by Colonel Charles Halt:

Col. Charles Halt made various statements about the incident in the years after it occurred. In his statements, he repeatedly emphasized that he and his security personnel were confronted with the unusual events and that they could find no conventional explanation for them. He claimed that the UFO on the ground was emitting unusual radiation and that they had tried to document the incident. The observations and experiences of Col. Charles Halt and his security personnel were instrumental in recording and investigating the incident. Halt also expressed the opinion that the US and British governments had downplayed the incident and failed to investigate it properly. He accused the authorities of a cover-up and lack of transparency. Their statements and documentation helped to make the Rendlesham Forest UFO incident one of the best documented and researched UFO cases.

Further investigations and interpretations:
Despite efforts by Halt and others to clarify the incident, there is still no single explanation. Some researchers and skeptics argue that the observed lights and metallic object were based on natural phenomena or secret military exercises. Others hold on to the extraterrestrial origin of the events and see the incident as proof of contact with extraterrestrial intelligence.

Significance:
The Rendlesham Forest UFO incident and the role of Colonel Charles Halt are of great significance to UFO research and public perception of UFOs. The incident has led to an intense debate about alien visitation, military cover-ups and government transparency. Col. Halt and his security personnel experienced the phenomenon firsthand and their accounts continue to raise questions that concern scientists and the public alike.

Incident of Ariel (1978)

The Ariel incident in 1978 is one of the most fascinating and puzzling incidents in the field of UFO sightings. It occurred near the city of Ariel in Israel and was witnessed by several eyewitnesses.

Event description:
 On the evening of January 21, 1978, numerous people in the Ariel region observed an unusual light phenomenon in the sky. Witnesses reported seeing a brightly glowing object moving in rapid and seemingly irregular movements. Some also described how the object changed shape and lit up in different colors.

Witness statements:
 Several eyewitnesses, including police officers, military personnel and civilians, gave accounts of the incident. Their statements largely coincided with regard to the observation of a bright light object that moved quickly and agilely. Some witnesses even claimed that the object hovered in the air for a short time and then disappeared at high speed.

Investigation and investigation:
 After Ariel's incident, investigations were conducted to find a possible explanation for the phenomenon. Government authorities and UFO researchers participated in the investigation to analyze the incident and verify witness

statements. Interviews were conducted, evidence was collected and data from radar recordings were analyzed.

Possible explanations:
There are various hypotheses and explanations for Ariel's incident. Some researchers suggest that it could be a natural phenomenology, such as an unusual atmospheric feature or a meteorite. Others, however, speculate on the possibility of an extraterrestrial origin based on the object's unusual movements and shape change.

Open questions and further research:
Despite the investigations and discussions, many questions remain unanswered. There remains a need for additional investigation and research to fully understand Ariel's incident. Additional witnesses should be interviewed, data and evidence analyzed in more detail, and possible technological explanations should be considered.

Conclusion:
The Ariel incident in 1978 remains a fascinating and unsolved UFO phenomenon. Testimony, investigations and discussions have shown that there are many different interpretations and hypotheses. A comprehensive and unbiased investigation of the incident, involving various disciplines and technologies, could help provide further insight and solve the mystery surrounding the Ariel incident. In order to conduct further investigation and research, it is important to verify the credibility of the witness

statements and explore possible technological explanations.

Varginha incident (1996)

The Varginha incident in 1996 is one of the best known but also one of the most controversial UFO incidents in Brazil.

Background:
On January 20, 1996, a series of events involving alleged alien sightings were reported in the Brazilian city of Varginha. Witnesses reported humanoid creatures with large heads and red eyes being spotted by local authorities and military personnel. The incident quickly gained national and international attention and led to intense discussion about the existence of aliens.

Witness statements:
Several witnesses, including police officers and civilians, claimed to have seen these alien beings. They described the creatures as being about 5 feet tall, with gray skin, large heads and red eyes. Some witnesses even claimed to have smelled an unusual smell that was associated with the creatures. The witness statements were recorded by media and led to widespread coverage of the incident.

Investigations:

Following the witness statements, Brazilian authorities opened an investigation into the incident. A team of military personnel, police and scientists was assembled to investigate the incident and find possible explanations. The investigations were shrouded in secrecy and controversy. Some authorities released official statements dismissing the incident as a misunderstanding or hysteria. Critics, however, claimed that evidence was covered up or manipulated to conceal the existence of aliens.

Controversies and Speculations:
The Varginha incident led to a lot of speculation and controversy. Proponents of the extraterrestrial hypothesis argued that the witness accounts and supposedly secret evidence pointed to an extraterrestrial visit. Skeptics, however, argued that it was a confusion of natural or conventional phenomena and that the testimonies were due to hysteria or misinterpretation.

Conclusion:
The Varginha incident remains a much discussed and controversial case. The testimonies and investigations suggest that something extraordinary happened in Varginha. However, there is no clear evidence of extraterrestrial activity. The incident remains a mystery and raises questions about the existence of extraterrestrial life and how governments respond to such incidents. The Varginha incident has also fueled the debate about government transparency regarding UFO phenomena and extraterrestrial li-

fe. To date, the Varginha incident has not been clearly clarified. Despite intensive investigations and witness statements, doubts and uncertainties remain. The truth behind the Varginha incident will likely continue to be a subject of speculation, controversy and debate.

Travis Walton Incident (1975)

The Travis Walton incident in 1975 is one of the most famous and controversial UFO abduction cases. Travis Walton claims to have been abducted by extraterrestrial beings, and his story has since garnered worldwide attention. This case study examines the events surrounding Travis Walton and analyzes the various viewpoints and theories that have emerged surrounding this incident.

The incidents:
 On November 5, 1975, Travis Walton was working with a group of woodworkers near Snowflake, Arizona. On the way home, they saw a glowing, flying object in the sky, which they interpreted as a UFO. Curious and fascinated by the sight, Walton decided to get closer to the object while his colleagues waited in the car. Suddenly, Walton was caught in a bright beam of light and seemingly abducted by a UFO.

The absence of Walton:
 After Travis Walton was kidnapped, his colleagues reported that he had disappeared. A nationwide search was

launched to find him. The police and media were involved and the incident received widespread attention. Five days later, Walton suddenly reappeared and claimed to have been abducted by extraterrestrial beings.

Travis Walton's experiences:
Travis Walton claimed that he was aboard an alien spacecraft while he was away. He described a series of medical examinations and interactions with the extraterrestrial beings. His story attracted a lot of attention and was hotly debated by both UFO enthusiasts and skeptics.

Here are some specific details he provided about these investigations: Walton claims that the aliens took samples from his body. This allegedly included blood samples as well as other tissue samples, the purpose and use of which were not clear to him. He describes that the aliens used strange instruments during the investigations, some of which were unknown to him. He describes these instruments as advanced and not comparable to the medical instruments he had seen on Earth. Walton claims that he was examined by humanoid beings, which he identified as extraterrestrial. He describes these creatures as tall and slender with large heads and large black eyes. They communicated primarily telepathically and did not speak to him in a human language. Walton states that his memories of the medical examinations are sometimes patchy and that he cannot remember all the details clearly. He describes how his memories of the events during the kidnapping became both clearer and blurrier over time. These descriptions of medical exams are part of Travis Walton's controversial narrative about his alleged alien abduction. They are seen by some as evidence of the reality of extraterrestrial visits to Earth, while others see them as part of a larger debate about UFO phenomena and human perception.

The controversies and theories:
 Travis Walton's incident sparked a controversial debate. Skeptics argue that Walton may have made up his kidnapping story, either for attention or to distract from a possible crime. Some also claim that it is a staged fake news.

On the other side are the proponents who believe Travis Walton and view his experience as a real alien abduction. They point to the consistent narrative and the fact that Walton and his colleagues were found credible during a series of lie detector tests.

The effects:
 Travis Walton's incident had a significant impact on his life and those of his colleagues. They were heavily criticized by the public and media and had to deal with the consequences of their story. The incident also impacted the UFO research community and sparked further discussion about alien abductions.

Conclusion:

The Travis Walton incident remains a mystery to this day and is one of the most famous UFO abduction stories. The study showed that there are many different points of view and theories regarding this incident. Skeptics question Travis Walton's credibility and accuse him of making up the story. Proponents, on the other hand, consider his experience to be real and see it as evidence of alien abductions. What is important is that, despite the controversial discussions and contradictory statements surrounding Travis Walton's incident, there is still no clear scientific confirmation, nor any rejection. Individual beliefs and personal interpretations play a large role in evaluating this case. Regardless, the Travis Walton incident helped raise public awareness of the issue of alien abductions and stimulate further discussion about the existence of extraterrestrial life. The case has influenced the UFO research community and led to further investigation and debate. Ultimately, the Travis Walton incident remains a fascinating chapter in the history of UFO phenomena. It remains important to carry out scientific investigations and critical considerations in order to form an informed opinion and better understand the phenomenon. Further research and open exchange are crucial to finding answers to the questions surrounding alien abductions and potentially revealing the truth about such events.

Betty and Barney Hill (1961)

The Betty and Barney Hill incident in 1961 is one of the most famous and investigated cases of suspected alien abduction. Married couple Betty and Barney Hill claimed to have been abducted and examined by aliens while driving. The incident attracted worldwide attention and led to intense investigation into the alien abduction phenomenon.

Background:

Betty and Barney Hill were an African-American couple returning from a vacation in Canada in the early hours of September 20, 1961. While driving through New Hampshire, they noticed a strange glowing object in the sky following them and arousing their fears. When they returned home later, they discovered that they had lost several hours without any memory of what had happened.

The incident:

After the incident, Betty and Barney Hill initially only had fragments of memories of strange encounters with strange beings. They then sought help from a psychiatrist who conducted sessions under hypnosis to restore their memories. Under hypnosis, they reported an abduction by aliens, which they examined and subjected to medical procedures on board a spaceship.

A notable claim made by Betty Hill was that during her abduction she saw a kind of star map showing the place of origin of the alien beings. This map allegedly showed a series of stars and star systems that she claimed represented the hijackers' hometown.

Interestingly, Betty Hill claimed that this star system was only discovered by scientists years later, around 30 years after her abduction. She claimed that the discovery of this star system by the "Happel Telescope" confirmed what she had seen during her abduction.

However, there is no reliable evidence that Betty Hill's claims are actually true. There are also no records of the "Happel Telescope" discovering such a star system. Betty Hill's story is considered by many to be part of a broader discussion about UFO phenomena and human perception. Nevertheless, their story remains an important part of UFO history and is viewed by some as evidence of the existence of extraterrestrial intelligences on Earth.

Investigations and Impact:

The case of Betty and Barney Hill quickly attracted the attention of the media and UFO researchers. A thorough investigation was conducted, interviewing witnesses and collecting evidence. Although there was no clear physical evidence of the kidnapping, Betty and Barney Hill's accounts were considered credible by some.

The case also had an impact on public awareness of alien abductions. He helped popularize the idea that humans could be abducted and studied by aliens. The case of Betty and Barney Hill has been featured in numerous books, documentaries and films and has served as a template for further research and discussion on the alien abduction phenomenon.

Criticism and debates:

As with many alien abduction cases, Betty and Barney Hill has its critics and skeptics. Some argue that the memories under hypnosis could be suggestive and that the incident may be due to other causes such as sleep paralysis or hallucinations. Others accuse Betty and Barney Hill of making up the story to gain attention or financial gain.

Conclusion:

The case of Betty and Barney Hill remains a fascinating and controversial incident in the field of alien abductions. Although there is no definitive evidence to clearly confirm or refute the incident, it has had a significant impact on the study and discussion of the alien abduction phenomenon. The study showed how such incident reports can pique public interest and lead to further investigation. The couple's detailed stories have inspired many people to share their own experiences with alien abductions and search for answers. But the impact of the case goes beyond the phenomenon of alien abductions. He has also contributed to the debate about the credibility of eyewitness testimony, the role of hypnosis in memory recovery, and the existence of extraterrestrial life in general. It remains a challenge to scientifically investigate the case of Betty and Barney Hill and find clear evidence. Nevertheless, their story has helped raise awareness of the possibility of extraterrestrial life and extraordinary phenomena in society. The case of Betty and Barney Hill continues to be discussed and studied by UFO researchers, skeptics and enthusiasts. It remains an open question whether her abduction actually took place or whether it was a case of faulty memory or other explanations. But regardless of the actual nature of the incident, it sparked interest in extraterrestrial life and the unknown and left a lasting mark on the history of UFO research.

Shag Harbor incident (Canada, 1967)

The Shag Harbor incident in 1967 is one of the most famous UFO incidents in Canada. It concerns an incident in which an unidentified flying object is reported to have crashed in the waters of Shag Harbor. This incident not only caught the public's attention, but also prompted the Government of Canada to launch an official investigation.

Background:
On the evening of October 4, 1967, several witnesses near Shag Harbor observed a large, glowing object falling from the sky and eventually disappearing into the waters. Numerous calls were made to the local police and coast guard, who then launched a rescue operation as it was initially assumed that a plane had crashed.

Authorities investigation and response:

The Royal Canadian Mounted Police (RCMP) and the Canadian Coast Guard were involved in the Shag Harbor incident. They immediately launched a search and rescue operation, using divers, boats and floodlights to locate the supposedly crashed aircraft and rescue survivors. However, despite intensive efforts, no wreckage or survivors could be found.

Official investigation:
The Government of Canada responded to the Shag Harbor incident by launching an official investigation. A task force was formed to investigate the incident and find possible explanations. The task force included representatives from the RCMP, Coast Guard, Department of Defense and other government agencies.

Witness statements were collected and analyzed during the investigation. Several witnesses reported seeing an unusual object fall into the water and then disappear without a trace. There were also reports of strange lights and noises near the incident site. Authorities also investigated other possible causes, such as plane crashes or military activity, but were unable to find a convincing explanation.

Result and aftermath:
The official investigation into the Shag Harbor incident ultimately yielded no concrete results. The government issued a press statement confirming that something unusual

had occurred at Shag Harbor but was unable to provide a clear explanation.

The Shag Harbor incident attracted both national and international attention. It received extensive media coverage and helped further increase interest in UFOs and extraterrestrial life. The incident is still discussed and researched today by UFO researchers, enthusiasts and skeptics.

Some witnesses to the Shag Harbor incident claimed that the crashed object could have been an alien spacecraft. They reported unusual lights, floating objects and an inexplicable silence in the area after the incident. These reports added to speculation that this was an encounter involving alien technology.

Despite the lack of concrete evidence and official investigation, the Shag Harbor incident remains a mysterious event that still raises many questions to this day. UFO researchers and interested parties continue their efforts to find more information and possible explanations for the incident.

The Shag Harbor incident also helped increase awareness of UFO phenomena and extraterrestrial life. He has stimulated public interest and contributed to the discussion about the existence of extraterrestrial civilizations. The incident is regularly covered in books, documentaries and

UFO conferences and is considered a significant chapter in the history of UFO research.

Conclusion:

The Shag Harbor incident is also an extraordinary testimony in UFO research. Although no definitive explanation has been found, the witness statements and the official investigation continue to raise questions and maintain interest in UFOs and extraterrestrial life. The Shag Harbor incident stands as an example of a well-documented and investigated UFO incident that continues to fuel curiosity and the pursuit of answers to the unsolved mystery of extraterrestrial life.

Cash-Landrum incident (USA, 1980)

The Cash-Landrum incident is a notable event that occurred in 1980 in the state of Texas, USA. It is a UFO incident in which several witnesses had an encounter with an unidentified flying object. The incident attracted widespread attention and continues to raise questions today.

Description of the incident:

On December 29, 1980, Betty Cash, Vickie Landrum and Colby Landrum were in a car returning from a restaurant near Huffman, Texas. Suddenly they noticed a bright light in the sky approaching them. The light was described as a large, diamond-shaped object that breathed fire and gave off intense pressure and heat.

The witnesses tried to avoid the object, but it appeared to be following them. After some time they found themselves near a road where they had to stop. They watched as the object landed nearby, attracting a large number of military vehicles and helicopters.

Betty Cash, Vickie Landrum and Colby Landrum witnessed an extraordinary scene where they saw an estimated 23 military helicopters and several military vehicles. The object was surrounded by strong flames and smoke. The witnesses were concerned about their health as they felt the radiation from the object was affecting them.

After the incident:
The witnesses suffered burns, nausea and other health problems due to suspected radiation exposure during the incident. Betty Cash even had to be hospitalized. Authorities and the military were informed of the incident and an official investigation was launched.

However, the official investigation did not provide a clear explanation for the incident. The military denied any involvement in the incident and claimed that no military operations were taking place in the area. However, there were reports of a secret military exercise near the incident site at the time.

Discussion and interpretation:

The Cash-Landrum incident remains a controversial event in UFO research. Some experts and UFO researchers believe it was an encounter with an alien spacecraft, while others are skeptical and seek alternative explanations.

One theory is that it may have been a secret military technology that appeared to witnesses to be a UFO. Another theory is that it was a combination of natural phenomena and mass hysterical perception.

Conclusion:
The Cash-Landrum incident remains an unsolved mystery in UFO research. Although official investigations have not found a clear explanation for the incident, witnesses continue to struggle with the impact of the incident. Betty Cash and Vickie Landrum have suffered health problems long after the incident and claim their quality of life has been severely affected.

The Cash-Landrum incident has also attracted media and public attention. Numerous reports were published and interviews were conducted with the witnesses. The incident has helped raise awareness of the issue of UFOs and extraterrestrial life and has further fueled the debate about whether we are being visited by intelligent extraterrestrial life forms. The effects of the Cash-Landrum incident are still felt today. He has helped promote interest in UFO investigations and the exploration of the universe. The incident has also prompted witnesses to speak publicly about their experiences and dedicate their lives to researching the phenomenon. Overall, the Cash-Landrum incident remains a fascinating and mysterious event that continues to capture people's imaginations and raise questions about extraterrestrial life and possible encounters with it. It remains to be seen whether future investigations and findings will provide more clarity about this incident and similar events.

Vostok Incident (Russia, 1982)

The Vostok incident in 1982 is a notable event in UFO research and has caused a stir in Russia and internationally. The incident occurred near Lake Vostok in the Eastern Siberia region and involves reports from witnesses of an unusual encounter with an unidentified flying object. This study examines the details and background of the Vostok incident and its impact on UFO research.

Background:

In 1982, three Soviet geologists were on an expedition near Lake Vostok. As they worked, they observed a bright light in the sky rapidly approaching them. The light was described by witnesses as a large, cigar-shaped flying object that hovered silently and was impressive in size.

The incident:

The witnesses reported that the flying object landed in the immediate vicinity and that there was exceptional silence. They described how several smaller objects became separated from the main object and moved in different directions. The geologists reported an intense feeling of awe and were surrounded by a kind of energy that they described as electrifying. During the incident, geologists also observed some sort of communication or interaction between the smaller objects and the main object. No aggressive actions or hostile intentions were detected on the part of the unknown flying objects. After a while, the smaller objects rose again and merged with the main object, which then quickly moved away.

Effects:

The Vostok incident had a significant impact on UFO research in Russia and worldwide. Witnesses to the incident reported their observations to the relevant authorities, but there is no official public statement regarding the incident. Nevertheless, reports and rumors about the Vostok incident spread among UFO enthusiasts and the public. The Vostok incident helped increase interest in extraterrestrial life and UFO apparitions in Russia. There has been a wide discussion about the phenomenon and various theories and hypotheses have been proposed to explain the incident. Some speculated about a possible extraterrestrial presence, while others considered alternative explanations, such as secret military experiments.

Conclusion:

The Vostok incident remains an unsolved mystery in UFO research. The witnesses' reports about the unusual encounter with the unknown flying objects in 1982 continue to raise questions and encourage further investigation. The Vostok incident draws parallels to other UFO encounters around the world and adds to the growing body of evidence and reports of unidentified flying objects.

To further explore the Vostok incident, future investigations could include the following steps:

Gathering Witness Statements: Detailed interviewing of the geologists who observed the incident is critical. It is important to accurately document their experiences, perceptions and emotions surrounding the incident.

Forensic investigations: Examinations of the environment in which the incident occurred could provide evidence of physical traces or alterations. Soil samples, air and water contaminants, and other possible effects should be analyzed.

Collaboration with other institutions: It would be useful to collaborate with international UFO research organizations and scientific institutions to leverage expertise and resources and look at the Vostok incident from different perspectives.

Technical Analysis: A thorough analysis of available photographic and/or video footage of the incident should be conducted. Identifying objects, trajectories, or other relevant features can help understand the incident in more detail.

Historical research: Comprehensive research into the region and any similar incidents in the past can provide further insight. It is important to consider historical records, accounts, or legends that may indicate similar phenomena. The case study of the Vostok incident highlights the complexities and challenges of investigating UFO apparitions. Although the Vostok incident raises many questions to this day, it helps raise awareness of the phenomenon of unidentified flying objects and continues to sustain curiosity and research. We can only hope that future research and advances in technology will help us solve the mystery of the Vostok Incident and similar occurrences and gain greater clarity on the nature of these phenomena.

Río Cuarto (Argentina, 1965)

The Río Cuarto incident in 1965 is one of the most famous and well-documented UFO incidents in Argentina.

Description of the incident:
On the evening of May 19, 1965, several residents of Río Cuarto witnessed an unusual event. They reported a glo-

wing, round formation in the sky that appeared in different colors and made strange movements. Some witnesses claimed that the object was accompanied by a bright beam of light.

Witness statements and investigations:
A thorough investigation was carried out to verify the credibility of the witness statements. Local authorities and scientists interviewed eyewitnesses and collected information about the incident. Witnesses described the UFO as large and bright, and some claimed that it remained in the sky for a short period of time before disappearing at high speed.

Physical Evidence:
Although no physical traces on the ground or other material evidence were found, the numerous witness statements are an important aspect in the investigation of the Río Cuarto incident. The witnesses' accounts are similar in many details, which indicates a certain level of credibility.

Discussion of possible explanations:
There are various theories that attempt to explain the Río Cuarto incident. One possible explanation is that it was a natural phenomenon such as a meteorite. Another theory is that it was a military experiment or secret technology. Another speculation is that it could have been an alien spacecraft.

Effects and Aftermath:

The Río Cuarto incident attracted the attention of not only the local population, but also the media and UFO researchers worldwide. The incident was covered extensively in the press and led to increased debate about the UFO phenomenon. It helped increase interest in UFO research and increase public awareness of unidentified flying objects.

Conclusion:

The Río Cuarto incident remains a mystery to this day, and there is no definitive explanation for the observed phenomenon. However, the numerous testimonies and careful investigation of the incident show that something inexplicable was happening in the skies of Río Cuarto. The incident adds to the growing collection of UFO reports and continues to prompt the scientific community and UFO researchers to investigate the phenomenon more closely. The Río Cuarto incident is another case that shows that there are many unresolved questions related to UFOs and that further research is needed to better understand these phenomena. Researching UFO incidents such as the Río Cuarto incident is important to gain more insight into possible extraterrestrial activity or unexplained phenomena. It is also of great importance to take witness statements seriously and conduct careful investigations to provide an accurate analysis of events. The Río Cuarto incident has had a lasting impact on people in the region and sparked their interest in UFOs and extraterrestrial life. It has also spurred discussion about the existence of extraterrestrial intelligence and the possibility of contacts with other civilizations. We must remember that the Río Cuarto incident is a significant event in the history of UFO research. Although no definitive explanation has been found, such case studies help expand our knowledge of UFO phenomena and focus attention on exploring the unknown.

Hessdalen lights (1983)

An unexplained phenomenon

Introduction: Hessdalen, a valley in Norway about 150 km south of Trondheim, is known worldwide for the so-called "Hessdalen lights" – mysterious light phenomena that have been sighted repeatedly since the early 1980s. These lights often appear at dusk or at night and move in the air as glowing spheres or pinpoints of light, often changing shape, size and brightness.

Description of the lights

Spherical: The Hessdalen lights often appear as glowing spheres that float or move in the air.

Color variations: The lights glow in different colors, such as white, yellow, red and blue.

Movement: They move in different patterns, from straight lines to abrupt changes of direction or random movements. Some appear to fly at high speed or disappear suddenly.

Investigations and hypotheses

Scientific theories to explain the Hessdalen lights

Ionization of the air: It is possible that the lights are caused by ionized air in combination with electromagnetic fields that are influenced by geological or meteorological conditions. It could be a kind of plasma.

Reflections or refraction: One hypothesis is that the lights are generated by reflections from vehicle lights or other natural light sources. However, as the lights also appear in remote areas without traffic routes, this theory is controversial.

Electrical phenomena: Some scientists see a connection with unusual electrical phenomena, similar to ball lightning, which could be generated by earth charges.

Atmospheric phenomena: Static electricity, magnetic storms or special meteorological conditions could also play a role.

Scientific and civilian research

Since the 1980s, scientists, the Norwegian military and physicists have investigated the phenomenon:

The Hessdalen Project (1983): this project has installed cameras and measuring devices in the region to document and scientifically analyze the lights.

Main scientists involved:

Dr. Erling Strand: as the project's leading researcher, he played a key role in setting up the measuring instruments.

Dr. Jørgen R. Isachsen: Physicist researching possible geophysical and atmospheric causes.

Prof. Håkan M. Svensson: Electrophysicist investigating electrodynamic effects as an explanation.

Reports and testimonies

Numerous eyewitness reports describe the lights differently, from very bright and focused to diffuse luminosity. The diverse descriptions make it difficult to form a uniform theory.

Tourism and interest

The mysterious lights have aroused the interest of tourists and ufologists, and a visitor center has been set up to provide information about the phenomenon. Many visitors hope to experience the mystery for themselves.

Link to UFO theories

Some speculate that the Hessdalen lights could be linked to UFOs. However, this theory is not supported by the scientific community, but has further fueled public interest in the phenomenon.

Hessdalen remains one of the most fascinating unexplained phenomena in the world. Despite intensive research and numerous hypotheses, there is as yet no conclusive explanation for the Hessdalen lights.

AATIP

The Advanced Aerospace Threat Identification Program (AATIP) was a secret US government program dedicated to research into unidentified aerospace phenomena. It was first announced in 2007 and aimed to analyze reports of unusual flying objects, also known as UFOs (Unidentified Flying Objects), and to identify potential security threats.

Background and objectives of the AATIP program:
The AATIP program was created by the US Department of Defense and was led by then intelligence official Luis Elizondo. The goal of the program was to better understand the nature and origins of unidentified aerospace phenomena and to identify potential threats to U.S. national security. The program was funded with an annual budget of approximately $22 million and worked closely with other government agencies, including the CIA and the Air Force.

Activities and methods of the AATIP program:
The AATIP program collected and analyzed reports of unusual flying objects reported by military personnel, pilots, astronauts and civilian witnesses. It relied on a multidisciplinary approach in which experts from various scientific fields such as aerospace engineering, physics, psychology and intelligence work worked together. The program used both publicly available information and se-

cret data sources to verify the reports and find possible explanations for the observed phenomena.

AATIP Program Results and Impact:

Although many details about the specific outcomes of the AATIP program remain classified, there are some known impacts that the program has had. One of the most significant was the release in 2017 of three videos created by AATIP that showed footage of unidentified flying objects. This publication caused a stir worldwide and led to increased public discussion on the topic of UFOs and extraterrestrial life. In addition, the AATIP program helped create awareness of the issue and promoted collaboration between various government agencies and scientific experts. It enabled the exchange of information and intelligence on unidentified flying objects and contributed to the topic being addressed more intensively.

Conclusion:

The Advanced Aerospace Threat Identification Program (AATIP) was a pioneering program to study unidentified aerospace phenomena. It helped raise public awareness of the issue and promoted collaboration between various government agencies and scientific experts. By analyzing reports and publishing videos, the program helped increase interest in UFOs and extraterrestrial life worldwide. Although the AATIP program was officially discontinued in 2012, interest in unidentified flying objects and extraterrestrial phenomena remains high. Continuing research in this area is being driven by independent organizations and private companies. The work of the AATIP program has laid the foundation for further investigations and shows that the topic of UFOs and extraterrestrial life remains of scientific and social interest. The AATIP program has not conclusively clarified whether the observed phenomena can actually be traced back to extraterrestrial intelligence. However, it has helped advance understanding and research in the field. Ultimately, the AATIP program has played an important role in the research and analysis of unidentified aerospace phenomena. It contributed to raising public awareness and promoting collaboration between government agencies and scientific experts. The results and impact of the program have increased interest in UFOs and extraterrestrial life worldwide and provided the basis for further research in this area.

This case study is based on publicly available information. Of course, certain aspects of the AATIP program remain secret.

Your own journey with aliens

Exchange with like-minded people

Exchanges with like-minded people and learning together are fundamental to the journey in the context of extraterrestrial contact. The need for connection and sharing experiences with others who share similar interests and experiences is at the heart of this fascinating journey and plays a crucial role in personal growth. Interactions with like-minded people provide a lively platform for exchanging experiences, insights and questions about the extraterrestrial phenomenon. These valuable encounters enable people to support each other, learn from each other and grow together. Connecting with like-minded people creates an atmosphere of understanding and support within a community of people interested in similar topics.

There are many ways to connect with like-minded people and obtain in-depth information about extraterrestrial contact. Online communities and forums offer the opportunity to connect with people from all over the world and immerse yourself in in-depth conversations. Regional events, regular meetings and organized groups offer the opportunity to make personal contacts and exchange ideas in a familiar environment. In addition, workshops and seminars offered by experts and experienced practitioners offer the opportunity to gain deeper insights and learn practical skills. Books and scientific literature are rich sources of information and provide material for stimulating discussions. Expert lectures, interviews and valuable information sources can also be used to deepen understanding and knowledge of extraterrestrial contact.

When interacting with like-minded people, it is crucial to maintain respect for different perspectives and opinions. Extraterrestrial contact is a fascinating but controversial topic that can provoke different views. It is therefore essential to promote an atmosphere of respectful dialogue in which different opinions can be heard and discussed without prejudice or judgment.

Maintaining privacy and confidentiality is of utmost importance as many people share personal experiences. Respectful handling of confidential information and respecting personal boundaries are fundamental principles of exchange. Although contact with like-minded people is a valuable resource, critical consideration is essential. Not all information shared is necessarily scientifically based or reliable. Reviewing information and using multiple sources helps make smart and informed decisions.

Exchanging ideas with like-minded people and learning together are valuable tools on your path to extraterrestrial contact. Through dialogue and collaboration with people who share similar interests and experiences, you can gain new perspectives, deepen your knowledge and make valuable connections. A supportive community can inspire and empower you to make the most of your extraterrestrial contact and continue to grow.

Interesting facts

Five recognized UFO signs

A UFO, short for "Unidentified Flying Object", refers to a flying object that cannot be immediately identified. Here are five signs that could indicate it's a UFO:

Unusual flight patterns:
UFOs are often associated with unusual flight patterns that differ from traditional airplanes or helicopters. They could show sudden changes in direction, extremely high speeds or movements that are not typical for terrestrial aircraft or drones.

Luminous apparitions:
Many UFO sightings involve luminous apparitions. These could appear as bright lights, beams of light, or changing patterns of light in the sky. The glow can vary, from glowing balls to changing colors.

Silent movement:
Another sign of a UFO is the silence that is often associated with their movements. Unlike airplanes or helicopters, which usually produce a noticeable level of noise, many UFO witnesses report silent movements of the objects.

High speed:

A UFO can move at inexplicably high speeds. These rapid movements can occur over long distances in a short period of time and cannot be replicated by man-made aircraft or drones.

These speeds 100% violate the laws of aerodynamics as we know them.

Several witnesses:

UFO sightings are often observed by several people at the same time. When multiple independent witnesses provide similar descriptions of the same object, the likelihood that it is something unexplained increases.

UFO does not necessarily mean that it is an extraterrestrial spacecraft, the term simply means that the observed object cannot be identified immediately, or at all.

UFO sightings and nuclear weapons

Another phenomenon following UFO sightings is that deactivations of anti-missiles or other security measures have occurred near military bases or facilities with nuclear weapons. These reports often come from military personnel or witnesses who claim that UFOs were able to influence or even disable the electronic systems of missiles. A particularly well-known example occurred during an incident in November 2010 near several US nuclear bases. Several eyewitnesses reported that unknown flying objects hovered over the facilities and were not only detected by the radar stations, but also caused the defense systems to be deactivated.

Such claims are often considered controversial and contentious. However, many of these reports are based on witness statements, unfortunately there is no scientific evidence or convincing evidence to support such cases. Official government statements on such events are usually cautious and do not confirm such influences from UFOs. There are various hypotheses as to why such reports might occur. Some believe that it could be a technical malfunction or misunderstanding, while others are convinced that UFOs are actually capable of interfering with the operation of military installations. It remains a controversial topic in UFO research, and there is no hard evidence to support such claims, nor any evidence to the contrary.

Of course, one should always treat such information with caution and demand available evidence. Alternative explanations are also repeatedly offered, such as failures, misinterpretations or secret military activities that could play a role here.

Discussions about UFOs and possible influences on military systems continue, and it is challenging to draw clear conclusions. Researchers and experts continue their efforts to find sound explanations for these reports and better understand the underlying causes.

Paul Hellyer

Paul Hellyer, a former high-ranking militarist and politician, has claimed that there is contact with aliens and that preparations are being made for Earth to join a galactic federation. This remarkable point of view was made by Hellyer even though he was already retired at the time and thus derived no personal gain or loss from this revelation. The question that arises in this context is why a former minister and politician should lie.

Paul Hellyer was born on August 6, 1923 in Waterford, Ontario, Canada, and he died on August 8, 2021. In Canada, he was a prominent politician who held various government offices and positions throughout his career. As a member of the Liberal Party of Canada, he was first elected in 1949 as a member of the House of Commons, the lower house of the Canadian Parliament. During his political career he held various ministerial posts, including Minister of Defense, Minister of Transport and Chairman of the Treasury Board.

After his retirement from active politics, Hellyer devoted himself intensively to the study of UFO phenomena, extraterrestrial life and conspiracy theories. His statements regarding governments worldwide's covert contact with extraterrestrial beings and humanity's use of extraterrestrial technology are extremely contentious, but still debated. They are viewed critically by many experts and scientists and are often not accepted. Unfortunately, it will forever remain a mystery why a former minister and politician, who was no longer in office at the time, made such explosive statements. A crucial point that we should always keep in mind is that the cover-up and secrecy of information about extraterrestrial encounters goes back more than 80 years. Given this fact, the idea of? ?a galactic federation at least does not seem completely unthinkable.

Hellyer has also claimed that he personally has evidence of extraterrestrial presence and even showed it to a reporter. There are reports that he claims to have photos of aliens and their spaceships and to have shown them to a journalist. According to him, these photos and other evidence are intended to prove the existence of extraterrestrial intelligence on Earth.

Unfortunately, because there is no verifiable evidence to support Paul Hellyer's claims, some critics accuse him of using his position and reputation to spread untrue claims. At this point it should be emphasized once again that he was no longer in active government service at this point, so he had no benefit from any statement. Nevertheless, his statements and claims remain a topic of intense discussion in the UFO and extraterrestrial research community.

Haim Eshed

Haim Eshed, was born in 1933. He is an Israeli visiting professor of aerospace at various space technology research institutions. A retired Israeli military intelligence brigadier general, Eshed led the space programs in the Israeli Ministry of Defense for nearly 30 years. He is also a former chairman of the Space Committee of the National Research Council and Development for the Ministry of Science, Technology and Space and a member of the Steering Committee of the Israel Space Agency. Eshed is often referred to as the father of Israel's space program.

Throughout his career, Eshed was awarded the Chief of Staff Citation, the highest non-combat honor bestowed by the IDF. He also received the Israel Defense Prize, the State of Israel's highest civilian defense award, three times, although the exact reasons for these honors remain secret.

Eshed served in the top-secret Unit 81, which provided technological solutions to the IDF's Military Intelligence Directorate. He holds a bachelor's degree in electrical engineering from Technion – Israel Institute of Technology and a master's degree and doctorate in aeronautical engineering.

In December 2020, Eshed made headlines when he claimed in an interview with Israel's national newspaper Yediot Aharonot that the United States government had been in contact with extraterrestrial life for years and had signed secret agreements with a "Galactic Federation." He also claimed that there is a joint underground base on

Mars where they work with American astronauts. These claims sparked a controversial discussion.

Parts of the interview were published in English by the Jerusalem Post, which led to further dissemination of the statements. A UFO investigator, Nick Pope, expressed doubts to NBC News about the credibility of Eshed's statements, asking whether it was primary or secondary information.

Eshed's book, "The Universe Beyond the Horizon: Conversations with Professor Haim Eshed," written by author Hagar Yanai and published in November 2020, contains further claims of extraterrestrial contact. Eshed tells stories about how aliens are said to have prevented potential nuclear disasters, including an unspecified nuclear incident during the Bay of Pigs invasion.

Isaac Ben-Israel, then chairman of the Israel Space Agency, expressed concerns to the Times of Israel about Eshed's claims, emphasizing that there were no signs of extraterrestrial life so far.

David Marler

David Marler is a well-known UFO researcher and author who specializes in the study of UFO sightings and unidentified flying objects. He is best known for his extensive research on the phenomenon of "black triangles", which are considered a special type of UFO.

Marler has written about UFO sightings in numerous books and articles, including "Triangular UFOs: An Estimate of the Situation" (2002) and "UFOs: Myths, Conspiracies, and Realities" (2017). He is often invited as an expert on UFO topics and has participated in various conferences and events.

During his time at MUFON, he had conducted numerous investigations into alleged UFO sightings and related experiences. He has discussed the topic of UFOs on numerous radio and television news programs. Over the years he has also given lectures on this topic to various schools and adults.

The UFO researcher has an extensive personal library of UFO books, journals, magazines, newspapers and case files from around the world, covering the last 70+ years. In doing so, he examined the detailed history of UFO sighting reports and associated patterns.

Marler received his Bachelor of Science in Psychology from Southern Illinois University at Edwardsville (SIUE). He received his certification in hypnotherapy from the Mottin and Johnson Institute of Hypnosis in St. Louis, Missouri. He is a Registered Polysomnography Technician

(RPSGT) and worked for several years at a large medical facility in St. Louis.

David is an independent UFO researcher who strives to be open-minded about the UFO phenomenon, but also recognizes the need for a skeptical approach when examining each UFO report.

Reputable researchers have known for years that triangular UFOs are among the most frequently observed types. The phenomenon has sparked heated debate among many and captured the imagination of many others. David Marler has provided a comprehensive analysis of the Triangles by collecting, compiling and analyzing hundreds of reports. His findings are documented in his book "Triangular UFOs: An Estimate of the Situation."

Lou Elizondo

Lou Elizondo and Chris Mellon are two personalities well known in connection with UFO research and revelations.

Luis "Lue" Elizondo, a former US military intelligence officer and UFO expert, became known for his role as head of the Advanced Aerospace Threat Identification Program (AATIP), a secret Pentagon program to study unidentified aerial phenomena. Born to a Cuban exile father, Elizondo grew up in Texas and graduated from Riverview

High School in Sarasota, Florida, before studying microbiology, immunology and parasitology. His military career included 20 years in the U.S. Army, during which he led intelligence operations in various parts of the world, including Afghanistan and Guantanamo Bay.

As head of AATIP from 2012, Elizondo was instrumental in studying unidentified aerial phenomena. His resignation in 2017, in protest against excessive Pentagon secrecy and internal opposition, made him one of the most prominent figures in the UFO discussion. Elizondo claims that the AATIP program existed from 2007 to 2012 and dealt with threats from space, including UFOs.

Elizondo joined "To the Stars Academy of Arts and Sciences" after his retirement and worked to bring information about UFOs to the public. He released videos of U.S. Navy UFO encounters that became known as the Pentagon UFO videos and helped produce the documentary "Unidentified: Inside America's UFO Investigation."

His claims and efforts have led to widespread discussion about the UFO phenomenon. Elizondo believes that UFOs may come from another dimension and that the US government may be in possession of "exotic material" related to these phenomena

Life:

Luis "Lue" Elizondo, a former US military officer and UFO expert, became known for his role as head of the Pentagon's Advanced Aerospace Threat Identification Program (AATIP). Elizondo was born in the USA and spent his youth in Texas. He graduated from Riverview High School in Sarasota, Florida, and went on to study microbiology, immunology and parasitology.

His military career spanned 20 years, during which he led intelligence operations in various parts of the world. As head of AATIP from 2012, Elizondo was instrumental in studying unidentified aerial phenomena. His resignation in 2017, in protest against Pentagon secrecy and internal opposition, made him a prominent figure in the UFO discussion.

Elizondo is a member of the To the Stars Academy of Arts and Sciences and is committed to bringing information about UFOs to the public. He was instrumental in releasing Pentagon UFO videos and co-produced the documentary "Unidentified: Inside America's UFO Investigation." His efforts have led to widespread discussion about the UFO phenomenon and the government's role in investigating UFOs.

Chris Mellon

Chris Mellon and Lou Elizondo are two prominent figures in the field of UFO research and disclosure. Lou Elizondo, a former U.S. Department of Defense intelligence officer, led the Advanced Aerospace Threat Identification Program (AATIP), dedicated to detecting and analyzing unidentified aerial phenomena, commonly known as UFOs. Following his time at AATIP, Elizondo became a strong advocate for the public release of UFO information and actively engaged in promoting discussions on the subject.

On the other hand, Chris Mellon, an experienced government official within the U.S. Department of Defense, also held positions with access to sensitive information. He later joined the UFO disclosure movement and became a member of the UFO research group To the Stars Academy of Arts & Science (TTSA). Both Elizondo and Mellon have emphasized the need for transparency in UFO reporting and research, aiming to enhance scientific investigation to gain a deeper understanding of the phenomenon.

While there is no definitive evidence that Chris Mellon had access to all security-related capabilities of the U.S. intelligence community, it is important to note the inherent secrecy and uncertainty surrounding UFO-related information. Public access to such activities and revelations is often limited.

Christopher Karl Mellon, born on October 2, 1957, has emerged as a key figure in the recent surge of UFO/UAP research. Coming from the influential Mellon family in Pittsburgh, his extensive political career included prestigious roles such as assistant secretary of defense for intelligence and staff director of the Senate Intelligence Committee in the Clinton and George W. Bush administrations. Mellon's academic achievements include a bachelor's degree in economics from Colby College and a master's degree in international relations from Yale University.

After working on Capitol Hill and various Pentagon roles, Mellon transitioned into lobbying and national security consulting before delving into UFO/UAP phenomena research and disclosure. Over time, he shifted from initial skepticism to a deep interest in investigating and disclosing these phenomena. Instrumental in the release of Pentagon UFO videos, Mellon has actively engaged in media appearances and documentaries to share his perspectives and raise public awareness of the UFO/UAP issue.

John Callahan

John Callahan, former head of the accident investigation division of the Federal Aviation Administration (FAA) in the United States, played a crucial role in a remarkable incident that occurred in November 1986. As an experienced aviation expert, Callahan had the responsibility to investigate and analyze aviation accidents and incidents.

At the center of the action was an incident of extraordinary UFO sightings reported by the crew of a Japanese aircraft during their transpacific flight.

During a flight to Japan, the crew of a Boeing 747 reported an unidentified flying object (UFO). FAA officer John Callahan led the investigation into the incident. After a thorough analysis of the radar and voice recordings, he concluded that the UFO represented a truly unexplained phenomenon.

Callahan organized a press conference at which he showed the radar images and audio recordings. At this conference he said that the UFO was real material and that it was visible to everyone. He also explained that this is a global phenomenon and not just a product of US air defense systems.

A quote from John Callahan is: "We were never here, this meeting never happened." This quote refers to Callahan's description of senior government officials attempting to cover up the incident. These statements underscore the alleged efforts to keep the incident secret. This incident remains unexplained to this day, and the exact nature of the observed object is unknown.

John Callahan was directly involved in the investigation of this incident because, as head of the Accident Investigation Division, he was responsible for evaluating unusual airspace incidents.

The incident included several visual observations and radar recordings of unidentified flying objects accompanying the Japanese aircraft. These inexplicable phenomena caused considerable concern and required thorough analysis. Callahan conducted the investigation with due seriousness, using standard FAA protocols.

What made the incident particularly unique was not only the quality of the observations, but also the subsequent handling of the information. Callahan decided not to cover up the results of the investigation but to make them available to the public. This unconventional transparency regarding UFO sightings by a high-ranking government official was unusual and generated considerable interest.

The investigation led by Callahan and his decision to release information have significantly increased the discussion about UFO sightings and the government's stance on them. Through his commitment to disclosing information and his contribution to this incident, he has established himself as a significant figure in UFO research.

Linda Moulton Howe

Linda Moulton Howe, a renowned investigative journalist and documentary filmmaker, has earned a distinguished reputation for her in-depth work in the field of alternative science, particularly regarding UFOs, aliens, cattle mutilations and other paranormal phenomena. Her career has been marked by dedicated research and a tireless pursuit of truth in areas often shrouded in secrecy and controversy.

Born on January 20, 1942, Linda Moulton Howe began her career as a TV producer and reporter. Her passion for investigative journalism led her to venture into areas that many of her colleagues shied away from. She was particularly known for her pioneering work in UFO research and paranormal phenomena. Linda Moulton Howe's groundbreaking work spans several decades, and her contributions have significantly influenced the way people think about these mysterious areas of science.

Howe found an early interest in phenomena that went beyond mainstream journalism. Her curiosity and daring led her to research and report on UFO sightings, alien contacts and livestock mutilations, which were often laughed at or ignored by the scientific community and the general public at the time. In the 1980s, she pioneered the TV documentary series UFO Report: Sightings. This series became internationally known and offered an in-depth look at various UFO sightings and phenomena. Howe's expertise and ability to explain complex topics in an understandable way helped make the series a success.

As an author, Linda Moulton Howe has published several books dealing with UFOs, alien contacts and other paranormal phenomena. These books are not only informative, but also reflect her ability to make complex topics accessible to a wider audience. Howe is known for taking a balanced approach to her work. Despite the sensationalism often associated with these topics, she has endeavored to provide reputable and accurate information. Her commitment to facts and critical thinking have made her a respected figure in alternative science.

Linda Moulton Howe's continued work, her lectures and her involvement in various media projects continue to raise awareness of the possibilities and mysteries of the unknown. Her insights and research have helped advance the discourse on UFOs and paranormal phenomena.

A milestone in her career was her extensive investigation of livestock mutilations in the late 1970s. Her in-depth reporting on this puzzling phenomenon brought her to the forefront of alternative science. Howe interviewed witnesses, farmers and experts to document the mysterious events and present accurate information to the public.

Linda Moulton Howe remains a key figure in alternative science, and her contributions have helped raise awareness of and promote dialogue about paranormal phenomena.

Bob Lazar

Bob Lazar is a controversial figure who claims that he worked on reverse engineering alien spacecraft at the secret base S-4 near Area 51 in the late 1980s. His claims have sparked great interest in the UFO community and led to much speculation about extraterrestrial technology and secrecy by government agencies.

Lazar claims that he examined nine different flying machines, which he referred to as "sports models", and that these spacecraft were powered by element 115, which had not yet been synthesized on Earth at the time. According to him, he played a key role in identifying and analyzing this alien technology, which he described as groundbreaking and revolutionary. Lazar claims that the craft he worked on used advanced technologies such as antigravity drives and space-time distortion to move through space. These claims have sparked both enthusiasm and skepticism in the scientific community and continue to be controversial.

Although Bob Lazar's claims have a large following, they are considered baseless and unbelievable by many scientists and skeptics. There are various contradictions and inconsistencies in his stories, and there is no verifiable evidence of his alleged activities in S-4 or of the existence of alien technology.

Robert Lazar, born to Albert Lazar and Phyllis Berliner, claims his resume was "erased" by secret government organizations. There is no public record of his education at Caltech or MIT, although he claims to have studied there. His statements about his professional career and personal life are contradictory and controversial.

Lazar's core message relates to his work on alien technology for the American government, particularly research into element 115 as a power source for alien spacecraft. These statements have been criticized and questioned by some scientists because they contradict current scientific knowledge.

Despite the controversy, Bob Lazar remains a well-known figure in UFO research and has a large following that continues to support and defend his stories.

David Grusch US Secret Service employee

US intelligence officer David Grusch claimed under oath before lawmakers in Washington in 2023 that the US government was in possession of crashed UFOs and even dead aliens. Grusch also stated in an interview that the Pentagon is in possession of UFO wreckage the size of a football field, which represents the remains of crashed UFOs that were definitely not man-made. Another point is that the Ministry of Defense should also have non-human biological preparations.

 With this statement made under oath against the Pentagon, David Grusch has clearly undermined the government's credibility. In this statement, the former intelligence officer and highly decorated Afghanistan veteran accused the Pentagon of, among other things, a cover-up. Grusch, a former Pentagon official who oversaw research into UFOs and unidentified anomalous phenomena, confirmed these statements under oath. Since the content in this case was strictly confidential, he also made it clear that from this point on he would fear for his life. He also suggested that there was a secret depot of extraterrestrial evidence somewhere in a basement at the Pentagon or some other agency. In his further statements, the whistleblower confirmed that it had never been made public that a bell-like UFO had crashed in Italy in 1933. This UFO was recovered by US soldiers during World War II and allegedly brought to the USA. According to Grusch, discussions were held with 40 eyewitnesses.

Former Navy pilot David Fravor also reported an encounter with an unidentified object in 2004. According to radar measurements, this object flew away at a speed of 5,760 km/h. Of course, these UFO encounters were also denied by the Pentagon.

The testimonies of David Grusch and David Fravor, made under oath, have now been made public.

Senator Harry Reid

Senator Harry Reid was an outstanding politician in the United States and held the position of Senate Majority Leader. His political career spanned several decades and he had a significant influence on US politics. In addition to his political achievements, Reid is also known for his role in UFO research.

Reid was born on December 2, 1939 in Searchlight, Nevada. He graduated from Utah State University and then served in the U.S. Army before entering politics. His political career began in the early 1960s when he became a member of the Nevada State Assembly. He later became chairman of the Nevada Gaming Commission before being elected to the U.S. Senate in 1986. Reid rose quickly in the Senate, eventually becoming majority leader of the

Democratic Party. This position gave him considerable influence and power in the US political landscape. During his tenure as Majority Leader, he played a critical role in establishing and funding the Advanced Aerospace Threat Identification Program (AATIP). This program, active from 2007 to 2012, aimed to collect, analyze and review reports of UFO sightings for possible security threats.

However, Harry Reid was not only known for his political achievements. His interest in unusual phenomena, especially UFOs, became public knowledge.

Reid played a key role in securing funding for the AATIP program. In this context, reports of UFO sightings were collected, analyzed and checked for possible security threats. The program lasted until 2012 and was then officially discontinued. However, it has helped raise awareness of the UFO phenomenon at the highest political levels. Reid's involvement in UFO research became public knowledge in 2017 when he appeared in a New York Times article reporting on AATIP and the U.S. government's research into UFOs. Reid stressed the importance of studying this issue seriously and called for a scientific approach to understanding UFO phenomena.

His attitude and influence helped to intensify the discussion about UFOs in the USA. Senator Harry Reid passed away on December 28, 2021, but his legacy in politics and his involvement in UFO research remains significant.

His decision to create and fund the Advanced Aerospace Threat Identification Program (AATIP) reflects a deep interest in research into UFOs and their potential threats in the United States Senate marked a milestone in the official discussion of the UFO phenomenon.

Reid's commitment to the AATIP was groundbreaking, as it was the first time official government resources and funding were dedicated to the systematic investigation of UFO sightings. This highlights his desire to apply scientific methods to a phenomenon that had previously often been relegated to realms of speculation and sensationalism.

The AATIP was created not only to document UFO sightings, but also to analyze possible security threats. This highlights Reid's pragmatic approach, which was based not only on scientific interest but also on the responsibility to understand potential threats to national security. The decision to fund the AATIP was certainly not without controversy, and Reid faced possible criticism. Nevertheless, this decision shows his willingness to go beyond political conventions and address issues that are often considered marginal.

Reid's moves to fund AATIP led to some normalization of discussion about UFOs in political and scientific circles.

The fact that a respected politician like Reid used the government's resources and influence to investigate UFO sightings gave the issue new credibility. It also opened the way for other researchers and experts interested in studying UFOs, paving the way for a rational, science-based discourse about these phenomena.

The consequences of Reid's involvement extend far beyond the sightings of unidentified flying objects. His contribution has helped open dialogue about the relationship between the government and potentially extraterrestrial phenomena. This bridging of politics and UFO research has not only drawn attention to the possibilities of extraterrestrial life, but also taken an important step towards transparency and openness in an area previously characterized by secrecy and speculation.

The impact of his involvement on public consciousness cannot be overemphasized. Through his leadership and support for AATIP, Reid brought the topic of UFOs out of the shadows of sensationalism and conspiracy theories and into the realm of serious research. His efforts helped raise awareness of the issue and stimulate widespread discussion about extraterrestrial phenomena in the United States.

General thoughts

The question arises in this context as to how far ahead of us possible extraterrestrial civilizations could be. Perhaps they have already reached such a high level of civilization that wars have long been a thing of the past for them. We could undoubtedly learn a lot from aliens, especially when it comes to humanity's biggest problem – limited resources versus growing populations. It is conceivable that extraterrestrials have already successfully found solutions to such challenges. It becomes clear that there are many things we do not yet know, many possibilities that lie ahead of us, and many risks that could face us. However, there are also many chances, especially considering how unlikely it is that we exist alone in the vast universe.

When we assume that something we don't understand is possible, it becomes magical for us. Magic is something we can't fight because we just don't understand it. That's why we tend to ignore it and banish it from our consciousness. There is so much we don't yet understand, and perhaps the reason is that we can't even imagine what there could be.

Statements from several former secret service employees provide additional cause for consideration. These people are not sensationalists, but experienced experts familiar with intelligence and security matters. Governments often claim that new flying objects are being tested to monitor the population's reaction. But it remains unclear why these flying objects are used in areas where there is no logical reason for them. There are dedicated areas to test new aircraft and it makes no sense to use them over densely populated areas or rural areas.

Some UFO sightings suggest that flight mode is designed as if someone were mapping, drawing and measuring the Earth. This claim comes not from casual observers, but from high-ranking military officials in the American armed forces as well as intelligence officials from the FBI and CIA. These are not people who may have seen something at some point; they are experienced observers with insider knowledge.

It's understandable why conspiracy theories exist. No matter what it is, high-ranking government officials, military leaders, and whoever, are helping to raise more and more questions. And that's a good thing, because why keep the population in the dark? Why do confidentiality clauses have to be signed? For many decades, apparently in the hope that the person who signed this confidentiality clause would die before it ended. If something falls to earth, it should be made available to the public and diligently informed about it. Then there is no need for a confidentiality clause.

However, I wonder why it should be so complicated since Fibonacci, a mathematical sequence, is used to try to establish communication or send information into space. In a discussion with friends, one idea was discussed again and again: beings that are able to fly over many kilometers, visit other planets, and reach the speed of light in the alleged 1000 km/h should not be able to to understand us? None of us believe in it. But what if beings from space have actually already visited our planet? Or even further, what if these aliens have been living among us for many years or centuries?

When, after many decades, high-ranking military personnel swear under oath that not only spaceships have landed on Earth, but aliens have also been found, that there are secret rooms underground, wherever they may be, in which several extraterrestrial flying objects are located, and research on flying objects would be carried out in these underground centers, and research would also be carried out on the bodies of fatally injured extraterrestrial creatures – this has already been announced several times. People who have been contractually bound to remain silent for more than five decades are now breaking that silence. And now is the right time to think about how we can communicate with aliens.

Experience reports from people who testify under oath, photos of UFOs taken in various places on earth – all of this should give us pause. And so the question remains whether they have been living with us for a long time or whether they are just visiting us. In any case, we should find some way to communicate with them. It is actually unimaginable that high-ranking military officers, researchers, people with degrees would take care of these UFO sightings. But only now going public with it. Nevertheless, it is still dismissed as nonsense and placed in the drawer of conspiracy theories.

People with normal minds will hopefully be able to use their brains to make up their own minds. Unfortunately, it is absolutely impossible to get documents that really say anything. But 50 years of secrecy also says a lot. Why are so many Air Force pilots and high-ranking officers afraid to speak when there are no UFOs anyway and they are all just weather balloons?

The title of the guide "Galactic Neighbors, Our Future with Aliens" gave us a lot in this book. But where there is no reading material, we can let our thoughts run free. Here we should take the opportunity to imagine why there is a reason to keep possible UFO sightings secret. Man should ask himself what would happen, what he would be capable of, if UFO sightings were publicly confirmed. If someone comes along and says that they have spoken to an alien and can prove it, the scientists have planned not to pass anything on. That it is expected that there will be mass panic. How would you or your neighbor react? Would everyone approach the extraterrestrial creature calmly, calmly and without fear? Or would people arm themselves to eliminate a perceived threat? This question must also be allowed and, above all, thought through in the context of communication with extraterrestrials. We were able to learn a lot with this guide and really thought about making these lines as simple as possible, but also as efficient as possible. We look forward to your thoughts and wish you interstellar and harmonious encounters and communications.

Epilogue

Dear readers,

With the conclusion of our guide on communication strategies with extraterrestrials, we would like to thank you for your attention and interest. We hope that you have gained new insights while reading and that this guide has further stimulated your curiosity about the topic of extraterrestrial communication.

The topic of communication with extraterrestrials is still a wide field of speculation and hypotheses. Our knowledge and experience in this area is limited and unfortunately many questions remain unanswered. Nevertheless, we believe it is extremely worthwhile to think about and deal with these questions.

We humans are naturally curious and have the urge to explore the unknown. The possibility of communicating with extraterrestrial life opens up a fascinating world full of possibilities and challenges. There is room for speculation, but also for scientific investigation and exploration. It is our responsibility to deal with this issue responsibly. Communicating with extraterrestrials requires not only technical and linguistic skills, but also a deep understanding of ethics, morals and cross-cultural sensitivity. Addressing these questions can develop us as humanity and encourage us to look beyond our own horizons.

We would like to encourage you to further develop your own thoughts and ideas on the subject of extraterrestrial communication and share them with other people. The exchange of perspectives and the discussion of different approaches are the key to a deeper understanding and a joint exploration of this fascinating field.

Finally, we would like to thank you again for your time and interest. We hope this guide has provided you with some food for thought or inspiration. In the future, may we discover new ways to communicate with extraterrestrial life and expand our own humanity in the process.

We should actually be very happy to live in this time. We have left the era of dinosaurs behind us and have seen how peoples, countries and generations have developed. Unfortunately, we have had to experience wars far too often in our previous existence. But we can be grateful if we were not directly affected. In our society there are people who are engaged in research in various areas and scientists who have already researched far more than is accessible to us. What particularly fascinates us all personally: We live in a time in which a lot of things are already in the past and a lot more are in the present. But our future still lies ahead of us. Who knows what it has in store for us? Nevertheless, we are absolutely convinced that our future has already begun and will not only take place with us humans on earth.

Thank you very much for your company on this fascinating journey.

Best regards,
 Sissi Ram and team

Sources

Here you will find publicly accessible sources for the studies listed that contain information about the processes mentioned.

Project Blue Book
United States Air Force. (1955). Project Blue Book Special Report No. 14. Washington, DC: United States Air Force.
Ruppelt, E.J. (1956). The Report on Unidentified Flying Objects. New York: Doubleday.
National Archives. (n.d.). Project Blue Book.

The Majestic Twelve (MJ-12)
Book: "The Report on Unidentified Flying Objects" by Edward Ruppelt
Book: "Top Secret/Majic" by Stanton T. Friedman
Documentation: Book "Mirage Men: An Adventure into Paranoia, Espionage, Psychological Warfare, and UFOs" by Mark Pilkington
Book: "UFOs and Government: A Historical Inquiry" by Michael Swords, Robert Powell, et al.
Documentary: "The Secret"
https://de.wikibrief.org/wiki/Majestic_12

The Kecksburg UFO Incident Book: "The UFO Encyclopedia: The Phenomenon from the Beginning" by Jerome Clark

Book: "UFOs: Generals, Pilots, and Government Officials Go on the Record" by Leslie Kean

Online news articles and archives: original reports and testimonies

Rudloe Manow

https://en.wikipedia.org/wiki/RAF_Rudloe_Manor

Online research: "Rudloe Manor UFO Incident" or "Rudloe Manor UFO". Online articles, forum discussions and other sources of information

Online forums and communities

Documentaries and TV shows

The Malmstrong nuclear missile crisis

Documentaries and TV shows

Roswell incident

Books:

"The Roswell UFO Crash: What You Don't Want to Know" by Kal K. Korff

Witness to Roswell: Unmasking the Government's Biggest Cover-Up by Thomas J. Carey and Donald R. Schmitt

Roswell: The Ultimate Cold Case Closed by Thomas J. Carey and Donald R. Schmitt

Documentaries:

"Unsealed: Alien Files" (Season 1, Episode 1 – "Roswell: Top Secret") – A documentary series covering various UFO and alien topics.

"The Roswell UFO Incident" (1994) – A documentary that looks at the events in Roswell and the various theories surrounding them.

Web sources:

Roswell Daily Record, July 8, 1947:

Historical archives and UFO research sites

Phoenix Light

Documentaries:

"The Phoenix Lights" (2005)

Books:

The Phoenix Lights: A Skeptic's Discovery That We Are Not Alone by Lynne D. Kitei, MD – A book written by an eyewitness to the incident

Newspaper articles and archives:

Arizona Republic, March 18, 1997

Web sources:

Phoenix Lights Network

Rendlesham Forest incident

Book: "Left at East Gate: A First-Hand Account of the Rendlesham Forest UFO Incident, Its Cover-Up, and Investigation" by Larry Warren and Peter Robbins

Documentary: "Rendlesham UFO Incident" (2014)

Newspaper articles and archives

Web sources

Ariel Incident
Book: "Ariel: The Unbelievable True Story of the UFO Abduction of an Entire School in 1994" by Randall Nickerson
Documentary: "Ariel Phenomenon" (2019)
Newspaper articles and archives
Web sources

Varginha incident
Book: "Varginha – Toda a verdade" by Marco Antônio Petit
Documentary: "Varginha: The Roswell of Brazil" (2002).
Newspaper articles and archives
Web sources

Travis Walton incident
Book: "Fire in the Sky: The Walton Experience" by Travis Walton
Documentary: "Fire in the Sky" (1993)
Newspaper articles and archives
Interviews and documentaries

Betty & Barney Hill
Book: "The Interrupted Journey" by John G. Fuller
Documentary: "The UFO Incident" (1975)
Interviews and reports

UFO research institutions: Organizations such as the Mutual UFO Network (MUFON) and the Center for the Study of Extraterrestrial Intelligence (CSETI)

Shag Harbor incident
Book: "Dark Object: The World's Only Government-Documented UFO Crash" by Don Ledger and Chris Styles
Documentary: "UFO Files: Canada's Roswell"
UFO research institutions: Organizations such as the Mutual UFO Network (MUFON) and the Center for the Study of Extraterrestrial Intelligence (CSETI)
Contemporary news reports and articles from Canadian media as well as international reporting

Cash Landrum Incident
Book: "The Cash-Landrum UFO Incident" by John F. Schuessler
Documentaries: "UFO Hunters: Alien Fallout" and "Unsolved Mysteries: UFO."
UFO research organizations: The Mutual UFO Network (MUFON) and the Center for the Study of Extraterrestrial Intelligence (CSETI)
Contemporary News Reports: "The New York Times."

Vostok incident
Book: "UFOs and the National Security State: The Cover-Up Exposed, 1973-1991" by Richard Dolan

Rio Cuarto incident
Book: "UFOs and Nukes: Extraordinary Encounters at Nuclear Weapons Sites" by Robert Hastings.

Hessdalen lights:
Informationsquelle ist die offizielle Website hessdalen.org, Journal of Scientific Exploration, Norwegian University of Science and Technology (NTNU), National Geographic

Advanced Aerospace Threat Identification
As far as we know, there is no specific case study titled Advanced Aerospace Threat Identification. However, it appears to be a topic that is being discussed in the context of UFO research and potential threats in the airspace. If you are looking for more information on this topic, you can search scientific articles, government documents or books on UFO research, air defense and military affairs. It is possible that information on this topic could appear under various terms or in connection with well-known organizations such as the US Department of Defense or the US Navy.

Paul Hellyer:
Website: https://www.paulhellyerweb.com/
https://en.wikipedia.org/wiki/Paul_Hellyer
Books: Hellyer, Paul. "The Money Mafia: A World in Crisis." Trine-Tag, 2014.

Haim Eshed:

https://en.wikipedia.org/wiki/Haim_Eshed

Interview in the Israeli newspaper Yediot Aharonot: https://www.ynetnews.com/article/BJQ7C8iIQ

Book: "The Universe Beyond the Horizon: Conversations with Professor Haim Eshed" by Hagar Yanai

David Marler:

Website: https://davidmarlerufo.com/

Book: Marler, David. "Triangular UFOs: An Assessment of the Situation." Richard Dolan Press, 2013.

Lou Elizondo:

Interview in the New York Times: https://www.nytimes.com/2017/12/16/us/politics/pentagon-program-ufo-harry-reid.html

Chris Mellon:

Article in Politico: https://www.politico.com/story/2019/06/01/ufo-sightings-navy-pilots-1340920

John Callahan:

Interview on the History Channel series "Unidentified": https://www.history.com/shows/unidentified-inside-americas-ufo-investigation

Linda Moulton Howe:

Website: https://www.earthfiles.com/

Book: Howe, Linda Moulton. "Glimpses of Other Realities: Volume II: High Strangeness." LMH Productions, 1998.

Bob Lazar:

Documentary: "Bob Lazar: Area 51 & Flying Saucers" (2018) by Jeremy Kenyon Lock

Interview in the Joe Rogan Experience Podcast: https://www.youtube.com/watch?v=BEWz4SXfyCQ

David Grush:

Congress hearing public 2023 / media

Harry Reid:

Interview in the New York Times: https://www.nytimes.com/2017/12/16/us/politics/pentagon-program-ufo-harry-reid.html

Disclaimer

If the publication contains links to third-party websites, the author accepts no liability for their content. Reference is only made to their content at the time of publication. All rights reserved. The use of parts of the book or even excerpts thereof without the author's permission is contrary to copyright law.

All information in this book has been carefully researched and checked; all information in this book is provided without any guarantee or warranty on the part of the author. The author accepts no liability for personal injury, damage to property or financial loss.

If you are satisfied with this book, please recommend it and the author to others. Please rate this book after your online purchase.

Picture credits / research:

All third-party photos, including the cover photo, were generated and researched with source credits or AI support.

Imprint:

Sigrid Trieb
Werk VI Strasse 22
A-8605 Kapfenberg

www.ingramcontent.com/pod-product-compliance
Lightning Source LLC
Chambersburg PA
CBHW071635220526
45469CB00002B/626